BY C. BROOKE WORTH

A Naturalist in Trinidad

Mosquito Safari: A Naturalist in Southern Africa

MOSQUITO
SAFARI

A NATURALIST
IN SOUTHERN AFRICA

BY C. BROOKE WORTH

SIMON AND SCHUSTER | NEW YORK

First printing

SBN 671-20827-6
Library of Congress Catalog Card Number: 77-139670
Designed by Edith Fowler
Manufactured in the United States of America

CONTENTS

Map on page 6
Illustration Section follows page 96

Countries and Locations Visited by
C. Brooke Worth in

MOSQUITO SAFARI

1

GRAND CENTRAL STATION

I could feel his eyes piercing me. But where *was* he? I had never seen Doctor Kenneth C. Smithburn in my life, and the information kiosk at Grand Central Station seemed an unusual place to have arranged a meeting, what with hundreds of strangers milling about. Why not the Bronx Zoo, in front of the giraffe enclosure, for example? At the exact stroke of noon there would not have been more than half a dozen spectators on hand, if that many, and a few polite inquiries would quickly have eliminated those gawkers who were not Dr. Smithburn.

The sensation of being transfixed persisted until my gaze was suddenly held by a pair of blue eyes belonging to a neat-looking, middle-aged, rather short gentleman with wavy white silken hair. Having "hooked a fish," the gentleman strode toward me, extended his hand and asserted, rather than questioned, "Brooke Worth."

That was one of the many things I always liked about Ken—his positiveness. For "Ken" it was at once.

"Let's get some lunch right away," he said. "I'm hungry, but I don't want to walk far, because I get short of breath. Will the restaurant here in the station suit you?"

Any place would have suited me. My future for several years

was contingent on the outcome of this meeting. After serving the Rockefeller Foundation in Florida and India for a number of years, I had left that organization's employ to work in Philadelphia on a public health project administered jointly by the University of Pennsylvania and the Wistar Institute. Four years of that routine had so surfeited me with city streets and back alleys that I was ready to commit murder, if such an exorbitant act would once again yield me a job outdoors in the tropics.

I had therefore written to Doctor Robert S. Morison, Director of Medical and Natural Sciences of the Rockefeller Foundation, stating my desire to be reinstated as a field staff member in any capacity then available. He replied that it just so happened that they had a current vacancy for an entomologist in the Union (not yet the Republic) of South Africa. Would I be interested in joining a team of virologists in Johannesburg as a mosquito expert?

I had never looked at an African mosquito, but I answered "Yes!" without mentioning that detail. "Very well," wrote Dr. Morison in his next letter, "but I want you first to meet Dr. Smithburn, who happens to be in New York on leave. He is the director of our research unit in Johannesburg, and it will be up to him to decide whether or not you qualify."

I had a million questions to ask Ken. For example, how could a mosquito-borne virus research project possibly be operating in Africa at the moment *without* an entomologist?

"Do you care for a cocktail before lunch?" he queried.

My response being enthusiastic, I went up the first notch in his estimation. From that time on, though we substituted food as nutriment after the third cocktail, nothing could break the rapport that had become established between us.

"When I was in Uganda," I found him saying, "I used to go out with Alexander Haddow on some of his monkey hunts. We were studying yellow fever, to determine the particular kinds of *Aedes* mosquitoes that transmit the disease at various locations: to monkeys in the jungle, among human beings in settlements, and intermediately between monkeys and people in fringe areas between forest and huts.

"One day Alec left me behind in his Land Rover while he went to look at some of his monkey traps. I was not particularly uneasy about remaining there alone until I heard a most ghastly animal sound of some sort advancing toward me. It was an almost indescribable gasping, coughing, gagging, grunting roar, and I was certain that a demented lion or gorilla was about to burst out of the bush and put an end to me. Soon a quite harmless and amiable baboon appeared, doing no more than clearing its throat."

"It's strange that we should have had the same experience," I replied. "In Honduras I once thought that I was being stalked by a jaguar. I ran away so fast that I never saw what was making those dreadful sounds, but when I reported the event back at our camp, my friends all laughed at me for having fled from howler monkeys."

"Isn't it extraordinary," said Ken, "how easily we are frightened outside our little houses of wood and brick."

"Well, I wouldn't call it entirely extraordinary," I said. "When you think of a howler monkey's equipment for making exceptional sounds—its inflated hyoid bone, connected with the trachea so that a reverberating chamber the size of a tennis ball can magnify—"

"Oh, so that's how they do it," interrupted Ken. "I have always heard about howlers but didn't know—"

"Coffee, gentlemen?" asked the waiter.

We ordered coffee, and then brandy, and got down to business.

During several decades of field and laboratory research on yellow fever, scientists of the Rockefeller Foundation had encountered a number of unidentified mosquito-borne viruses that were definitely not of the yellow fever sort, but which the urgency of yellow fever work did not give them time to study thoroughly. Those unknown viruses might be simply curiosities from the standpoint of human medicine. Perhaps they could injure only crocodiles or tapirs, and had gotten themselves into laboratory test tubes merely because they were residing within mosquitoes having extraordinarily catholic biting tastes. But now it was time to look into the mystery—to discover the natural history of those neglected viruses—and the Rockefeller Foundation had set up new programs in scattered parts of the world with the objective of sifting the global environment for as many new arboviruses as could be discovered—arboviruses being those viruses transmitted by arthropods, which are jointed-legged invertebrate animals such as insects, including mosquitoes and sandflies; arachnids, including ticks, mites and spiders; crustaceans, including crayfish, crabs and lobsters; and so on.

Doctor Smithburn had been the director of the yellow fever laboratory in Entebbe, Uganda. Under the new program he was assigned the task of organizing an arbovirus laboratory in Johannesburg in collaboration with, and drawing added support from, the South African Institute for Medical Research. South Africa had been chosen from the standpoint of its arboviral virginity—no such investigations had yet been made there—while the presence of an already sophisticated research organization and the universal use of the English language by scientists there were further attractions.

My qualifications as an entomologist rested on work done totally outside Africa. But since I had spent over two years as a mosquito man on one of the Foundation's malaria teams in India, it was conceivable that I might master the winged biting fauna of another continent. (Of course, African virus-carrying *culicine* mosquitoes were horses of far different colors from Indian malarial *anophelines,* but for the moment I did not dwell on that.)

It is quite a trick to look for invisible viruses, find them, and then tell the various kinds apart. Indeed, the trick seems to verge on magic, and certainly the mumbo jumbo of laboratory equipment, procedures and jargon would sound and look well in the highest-class sorcerer's kitchen.

But in reality the principles are straightforward, concerning mainly the biological needs for viral growth and multiplication. The lab is an arena for a bit of haphazard testing at first, since viruses often produce no visible changes in their natural hosts but make trouble in abnormal ones. The challenge then becomes one of finding a host that will betray a virus's presence by becoming sick or dying.

Surely no one can recognize sick mosquitoes. Infected ones look and behave just like uninfected ones. Therefore it is necessary to grind the mosquitoes up and inject suspensions of them into laboratory animals that will respond to whatever viruses may be there. Signs of infection may be tremors, paralysis or other derangements of the central nervous system, or there may be generalized hemorrhagic or other disturbances.

Viruses cause the formation of antibodies of several sorts. These may be compared in various ways to identify the viruses that called them forth. In the simplest case, portions of a suspension of virus to be identified are mixed in separate test tubes with sera containing known brands of antibody. The antibody in only one tube—that containing the kind of antibody called forth by the unknown virus—will react, neutralizing its virus. After the several mixtures are inoculated into animals, all the experimental hosts will get sick except the one that received neutralized, i.e., inactivated, virus. Thus the "unknown" can now be associated with its corresponding antiserum, and, the source of each being known, one can chalk up an additional example of the host-virus relationship.

From what Dr. Morison had written me, I gathered that several new arboviruses had already been discovered by Ken's team, but that the rate was now dwindling. However, up to the present time mosquitoes had been only a side issue with the part-time entomologists (whose specialties or responsibilities lay elsewhere). If I now became a dedicated 100 percent mosquito promoter, the level of adventure ought to rise, and everyone should begin having three times as much scientific fun.

My pledge to attempt that enhancement was all Ken Smithburn wanted to hear. All I wanted to hear from him were his words, "Africa, yes," because in my language that expression brought me tingling visions of a galaxy of wildlife in which mosquitoes were but one brilliant star. Indeed an arbovirus team could not afford to overlook any department of natural history, because wherever one peered one might find unsuspected viruses. Hence, where field work entailed in arbovirus investigations was concerned, versatile naturalists were sometimes more valuable than highly specialized ones. Every man who combined two disciplines saved the budget from having to embrace twin salaries; and he occasionally could link two lines of thought from his convenient perch on a bridge. Moreover there is much cross-fertilization among personnel working as closely together as the search for viruses demands, so that everyone eventually becomes a bit of an expert on mosquitoes, wild rodents, birds, big game, botany, geology, hydrology, anthropology and medicine.

All this in quest of not merely the invisible but actually the unknown! As Ken remarked, "We didn't know for sure that we'd find anything at all to start with. Our first South African expedition—to Simbu Pan in Tongaland—paid off far better than I could have hoped for. From a mere 7,637 mosquitoes we isolated seventeen virus strains belonging to six types of virus, of which three— Pongola, Simbu and Spondweni—were apparently new to science. And then our entomologists went off to do other things."

"What a time for intermission," I sympathized.

"Well, I guess that season must have been about over anyhow," Ken admitted. "But at least we now know that we are on to something good and it's worth chasing after viruses harder than ever. Our objectives—the Arbovirus Research Unit's objectives—are to find out how many such arthropod-borne viruses exist in the southern part of the African continent, what mosquitoes or ticks carry them, what hosts (including man) they infect, how common and how important each virus is, how it maintains itself and is spread, and what range it occupies. For all these investigations the lab depends on its field crew for collected samples—blood serum from man and other vertebrates in many instances, but, above all, mosquitoes. Don't forget: mosquitoes and more mosquitoes. Your job, Brooke, will be to identify mosquitoes until your eyes drop out."

2

OFF TO THE RACES

I kissed everyone goodbye twice on August 17, 1958, at the Philadelphia Airport—except for my wife, Merida, who missed the farewell gathering because she was ill at our South Jersey cottage. (I sent her flowers.) Auntie was there, and so were Valerie (great with our first grandchild) and George (responsible for that happy condition); Michael (accompanied by a current girl friend from the Wistar Institute); and Douglas (lightly moustached and singularly single). We had recently had a gala farewell night at the Admiral Hotel in Cape May, where Doug headed a jazz combo that summer, and George had led Val so violently in some sort of jitterbugging that Merida and I feared she might give birth right on the floor. Now the occasion was a bit more solemn, for I would not set foot again on American soil for two years. Nor would I see any of my family during that interval except for Merida's brief visit to Johannesburg.

Bob Morison had quickly seconded Ken's OK. My plane was a four-engined propeller affair, because the earlier jet Comets had begun to explode from metal fatigue, one by one, and travel facilities had regressed to questionable amenities of the Lindbergh era. That meant two nights and one day on this plane, as well as a problem related to food storage. I forget what airline it was, but their handsomely printed menu boasted that dinner the first night

13

had been provided by Maxim's in Paris. The lobster thermidor consequently had crossed the Atlantic once already, and by the time I ate it on the way to Lisbon it was prepared to attack such viscera as usually attend to digestion cooperatively if not insulted. I became deathly sick, which is a most inconvenient state on an airplane.

At Lisbon, early next morning, I was given a ticket that said I was entitled to a free "beverage" at the airport dining room. I elected to have a Portuguese brandy called Constantino, which I thought might buck me up. Wrong idea! That made things worse, and as we crossed the Sahara Desert during the forenoon, I almost hoped that we would crash so that I could expire. *In extremis*, I still could not keep my eyes from the window and dimly noted camel caravans crossing those wastes, exactly as recounted in my sixth-grade history book.

We reached Léopoldville, in the Belgian Congo, at some late hour of the night, and I struggled "ashore" simply to be able to say that I had set foot on the soil of that equatorial land. I have a lasting impression of native paintings for sale in the airport—watercolors in startling blues principally, but all of them somehow iridescent or metallic. Had I not been so groggy, I would have bought a dozen of them.

On that afternoon we had landed briefly at Accra, Ghana—a matter of so little consequence that I have almost forgotten to mention it. However, that stop afforded me an introductory glimpse of permanent significance. During our descent I noted so many fires that the entire continent seemed to be ablaze, and that impression remained throughout my next two years in southern Africa.

Ken Smithburn and Bruce McIntosh met me and were alarmed at my state. After shaking hands with my stocky, handsome, dark-haired new colleague, I said that all I wanted to do was go to a hotel and collapse. There seemed to be endless immigration formalities, but Ken attended to all of them except forging my signature here and there. Once in the car with my luggage, I sank back gratefully. However, a "birder" is incurably a birder as long as he can draw breath. As we were driving out of the terminal, I straightened up and said, "There's a wagtail of some sort on the lawn."

Bruce, who was the ornithologist (among his other capacities as serologist and veterinarian) on the virus team, looked at me with dawning appreciation. "Yes," he said, "it's a Cape Wagtail, peculiar to southern Africa." From that instant forth we became field companions, and he helped me immensely in getting oriented to the local avifauna, as well as showing me the best places to look for birds.

They took me to the Oxford Hotel in Rosebank, a fashionable suburban residential district of Johannesburg, where I carried out my threat and collapsed. I can't remember what happened during the rest of that day, but on the following morning I awoke more or less cured but excessively weak. Some pigeons outside my window were making a cooing racket which struck me as being unlike the things pigeons on City Hall in Philadelphia customarily utter. However, that might be due to activities of pigeon fanciers in South Africa, who had developed a special breed locally. I shut the birds out of consciousness, had some breakfast and then began to totter around the block in which my hotel was situated.

Almost at once I came to a bookstore. "Have you a book on South African birds?" I asked a white-haired female clerk.

"Indeed we have," she answered. "A very good one, too, by Dr. Austin Roberts, as revised by McLachlan and Liversidge only a year ago."

"How much is it?" I wanted to know. At the airport, Ken had helped me cash a traveler's check for fifty dollars or so, just to have money in my pocket. The lady now said that the book cost "forty and six" or some such figure.

"Oh, I'm afraid I haven't got that much with me," I said, taking out a wallet stuffed with pound notes. I would have that book, no matter how expensive it might be.

In the background stood another salesperson who, I found out afterwards, was the wife of a biochemist, Dr. John W. Hampton, who collaborated with our virus team. Norah never recovered from her amusement as the white-haired lady explained that "forty" meant shillings, not pounds, and that I had enough in my wallet to buy a dozen volumes. My clerk carefully extracted a few bills and then handed me back some change.

At the Oxford I sat down first of all to mark my name and the address of the lab on the cloth cover. This I always do with books that I take to the field, in case of loss. Then I began to go through the volume page by page. After an introductory section the text went on to penguins, grebes, albatrosses, and a number of other families, until herons occupied the scene. Then, suddenly, it began all over again with introduction, penguins, grebes, albatrosses, etc. I had bought an imperfect copy. Moreover, the duplicated pages were there in default of a large section covering other bird families. With that omission the book became virtually worthless.

Those damned pigeons continued their distracting cooing, and my convalescence went into a slump. I had crumpled the dust jacket and thrown it into a wastebasket (which I always do as well), defaced the cloth cover, and made a fool of myself in the shop. Did I dare go back to claim another copy? Well, yes—I did. The lady was extremely kind and made a substitution at once.

My first days in Johannesburg were occupied by activities having nothing to do with mosquitoes or any other advocates of viruses. Of course I had a look at the laboratory where I would work, but there was time for little more than an occasional glimpse of that modern spreading two-story concrete structure. For first I must buy a car and find a permanent place to live, since the lab was in the country, not within walking distance of any settlement. Everything took much longer than I would have thought necessary in this progressive place. I stayed at the Oxford for over two weeks, dependent on Bruce's kindness for driving me here and there.

Ultimately I got four wheels under me in a Renault Dauphine. Meanwhile my two feet had done a lot of promenading and I spotted the Garden Inn, only a block or two distant from the hotel and on a side street without heavy traffic. This establishment was well named, for it was actually a huge flower garden, with shrubs and trees interspersed along walkways that traversed lawns and floral borders. The Inn itself contained chiefly a large dining room, though a few people lived in quarters upstairs. Otherwise guests occupied separate small cottages on the grounds. These were constructed in several patterns, but the most typical was called a *rondavel,* "round hovel" in the Afrikaans language. Its general form was that of a brick cylinder with a conical thatched cap. The thatch was made of dried grass to a thickness of at least one foot.

Ken and Bruce, and other friends I had come to know at the lab by that time, approved my choice of a car, but they thought I was mad to be attracted by a rondavel. It was more fitting, they thought, that I should rent a house and hire a staff of servants. Failing that, I ought at least find a modern apartment and secure the services of a laundress and cleaning woman. But I *liked* the Garden Inn and the simplicity of a rondavel. I moved into a circular room that had no more than a bed, a table and chair, a washbasin with cold running water, and a tiny clothes closet. For two years that was my cloister, and everyone felt sorry for me but myself.

During the search I had conducted, my perambulations gave me a thorough familiarity with Rosebank. Though I always came back to the Garden Inn, in doubt about that haven only because of my friends' misgivings, I made a real effort to canvass all other possibilities. The region in general was unequivocally "right." Rosebank had a complete shopping center within its few commercial blocks that included grocery stores, a barber, a dentist, a "chemist's," department, hardware and liquor stores, and a movie house—to mention only some of the amenities. Immediately peripheral to that zone were several other boarding places and apartment houses, as well as the beginnings of a wide fringe of elegant residences.

On my last morning at the Oxford Hotel I was again awakened

by pigeons. By that time I had become convinced that there was something more then slightly odd about their voices. Indeed, I had watched them from the first day, when I felt close to eternity, and was even then impressed by the fact that they all looked alike, with red faces, but also stocky bodies and ruddy plumage generously spotted with white on the wings. I knew of no such domestic breed. Feral domestic pigeons, furthermore, are usually so genetically mixed that you can only rarely see two of matching patterns. They vary through black, brown, gray, fawn and white, some individuals being pied in all those shades. Often the distribution of colors is assymetrical, so that the wing of one bird will have several white flight feathers on the right, while the left is totally dark or light as the case may be.

At last I "came to" regarding a phenomenon that had confronted me since my first day in South Africa. These pigeons, as the excellent illustration in Roberts' book confirmed, were a wild species confined to Africa south of the Sahara wherever cliffs exist. After the white man came and erected cities, those birds regarded his buildings as so many additional escarpments and precipices, duly colonizing them. Their vernacular name, "Rock Pigeon," is a bit unfortunate, since our domestic pigeon is descended from the "Blue Rock Pigeon" of Europe and Asia. However, the similarity ends there, as I should have been quicker to recognize.

At the Garden Inn I was gratified to find that Rock Pigeons found other hospitable ledges on the main building. A thatched roof, no matter how thick, transmits sound admirably, even while shutting out rain and all but violent winds. Thus I continued to be awakened each morning by cooing sounds that sometimes seemed almost to be inside that little room. Now I liked them, since the birds were really native. In addition I heard another note that must come from a member of the same family. It was definitely not a "pigeon" sound. It did not take long to trace it to Cape Turtledoves in the overhanging acacia tree. Instead of cooing, the birds emitted a continuous syllabified utterance which Roberts has aptly rendered as "How's *fa*ther, how's *fa*ther," etc.

The acacia held hanging nests of Masked Weavers—almost a dozen such habitations, though that did not necessarily denote as many complete families. The males, yellow with black faces and throats, built nests interminably, as if they were the first do-it-yourself enthusiasts. Their industry, as well as their constant wheezing notes, attracted sparrowy-looking females which came to inspect the housing development as it progressed. Weavers are polygynous, a male usually claiming two or three wives, for each one of whom he builds a nest in succession. Females' choice of mates appears to be founded more on home appurtenances than on a spouse's charm, for all males display similar amorous courtship antics when females arrive, as far as a human being can discern, but those materialistic sluts brush past the panting swains to have a look indoors first.

Once a male weaver has acquired a mate and she has laid eggs, he abandons her in order to begin construction of a new domicile, not clandestinely at all, but right next door. That seems to be perfectly acceptable to Mamma Number 1, who now has a job that keeps her too busy to worry about bigamy. Accordingly, my acacia tree was full of nests, though I never saw more than five or six weavers in it.

Roberts' book informed me that the males make actually two sorts of nest—one for breeding and the other for sleeping. They were all wonderfully woven, for whatever function they had, suspended from tips of branches (from which all acacia leaves had been carefully snipped off, as in a well-tended hedge), and then fashioned into ovals out of pale, well-dried strips of grass and similar fiber. Breeding nests had tubular entrances, which the birds had to enter from below. For mere sleeping, males omitted that difficult portico after satisfying basic architectural specifications.

The grounds of the Inn supported several kinds of tree besides the acacia. Among these I was most impressed by English oaks, said to have been brought to South Africa by early British settlers in the form of acorns. That was a superlative scheme, whether the importers knew it or not, for by transporting only the germs of oaks across the sea they had left behind all the pests and parasites that normally riddle those trees in their native realm. I once read that oaks in the United States are hosts to more species of caterpillar than any other trees. So it may be in England, too. In any event, the oaks at the Garden Inn, though still not of conspicuous stature, were perfectly gorgeous in their foliage, with not a spot or wrinkle or ragged edge on any of their leaves. Apparently no native South African parasite was preadapted to attack oaks, and these trees gloried in an immunity that their British predecessors had never achieved.

I have never enjoyed thunder so much as I did in my rondavel. Johannesburg experienced seven months of constantly clear weather and five months of intermittent rain every year. At the onset of the rains, great electrical storms were likely to appear suddenly, flooding the streets so deeply that all the ladies, caught unawares, took off their shoes and stockings in order to wade home or to their parked cars in front of fashionable stores. It always seemed to me that we were actually inside the thunderclouds, and that may have been almost true, since Johannesburg is six thousand feet above sea level. In any case, the rondavel having a pervious thatched roof as far as noise is concerned, I would sit or lie, depending on the hour, and revel in claps and peals that literally engulfed or saturated the very air I was breathing.

The food at the Inn was excessively British but really quite good most of the time. Vegetables were overcooked, naturally. One, which I have not met elsewhere, was a spherical squash or pumpkin the size of a baseball, served boiled and whole. You were supposed to crack the egglike shell and scoop out and consume the interior, seeds as well as pulp—unless the seeds of a given squash were overripe; in that case, the pulp was still delicious. Plenty of fine beef was produced locally, so that steaks were commonly served. As by-products of that aspect of animal husbandry, oxtails and marrow bones on toast were splendid accouterments also. British? I should think so!

The Inn may have been run along British lines, but the management and guests alike were of mixed origin, some of the marrow-bone eaters being as purely Dutch as their cheese-consuming ancestors. Yet somehow or other the British tradition had triumphed over everything except architecture. Many of the inhabitants worked in central Johannesburg and commuted daily by car, while others were retired couples or solitaires who confided in the local pussycats with which the premises abounded. In keeping with the British aura, almost every one of those people maintained a façade of aloofness that, by the bye, suited me well, since I desired to be a solitaire too.

All the servants were Africans. Cooks, waiters, cleaners, house boys, gardeners: every one was dark-skinned. Racial intermixture had been taboo to such an extent that "coloureds" were exceptional, at least in Johannesburg. Therefore I believe that all the Garden Inn's employees were pure Negroes. Certainly the hue of their skins did not suggest any Caucasian contamination. As for their features, these southern African blacks were often much less typically Negroid in appearance than their more equatorial representatives, having, for example, almost aquiline rather than flattened noses in some instances.

I was greatly attracted to the structure, poise and efficiency of the waiter who catered to the table to which I was assigned. One thing that I noticed at once (and this was *not* one of the attractions) was that his earlobes had been pierced and stretched so that they now hung down as two unsightly flaps with inch-long holes in them. I presumed that he had grown up as a savage somewhere "in the bush" and had then been drawn to Johannesburg by the promise of improvement in his lot.

I was wrong about many things in that presumption. To take one: my waiter still took a savage's pride in his ears. On Sundays and holidays he always showed up with carved, brightly painted discs in his lobes. If he had not worn them periodically, I suppose the distended tissues would have begun to shrink, and then he would have been out of style.

One day, when he was thus ornamented, I naively ventured to ask him whether the plugs or spools—whatever they should be called—did not hurt when he put them in after days of not wearing them. He understood English quite well, never mistaking my orders and answering satisfactorily when I asked questions about items on the menu. However, he never smiled, and his eyes always had a distant, impersonal look in them. Now that I had asked him something intimate, he clammed up and a flash of hostility was apparent in his glance. From that time on he became more distant than previously.

The same was true of other servants at the Inn. The more friendly I tried to be, the further they retreated. Then I began to notice a reticence—really more an animosity—wherever I went. Whether it was only during a walk to the shops in Rosebank or an expedition to downtown Johannesburg, the Africans uniformly refused to return a greeting or even a smile. If two or three of them were approaching in gay native conversation, they would invariably interrupt their talk to put on mute, dour facial expressions as we passed. It was uncanny how they could deal matter-of-factly with required facets of their relationships to whites but would turn off their personalities like electric-light switches the moment those requirements had been satisfied.

Of course, I had come to the land of apartheid. Though I had read of that system, I did not realize until now how deeply it affected everyone in South Africa, both Negro and Caucasian. Our racial tensions in the United States are real enough, sad to say, but they can be lessened or nullified in individual cases when both parties want to be friendly. My public-health work in Philadelphia had taken me into countless Negro homes, in some of which I eventually became more happy and at ease than in many a college professor's stilted establishment. My ring at the door would bring cries of "Mamma, the doctor's here!" Then, while I asked warmhearted Mrs. Jones about her family's health during the past week, the smaller children would sit on the floor, clasping my legs affectionately.

I thought I would be able to get past apartheid in the same manner. But—at least in Johannesburg—there was some kind of generalized code that excluded individual exceptions. My friendly motivation made no dent whatsoever. When I had set forth to study mosquitoes on this continent, I felt I was "off to the races." Now I saw that expression in a new context.

Naturally I had taken a position against apartheid before coming to South Africa. But it is one thing to make such an intellectual judgment in a reflective and composed manner, and something else again to find oneself immersed in the situation itself. I was simply not prepared to realize how virulently the antagonisms engendered by apartheid tinctured every resident of South Africa. My first innocent questions elicited such violent answers that I soon stopped making further queries but became a sensitized listener and a chronically alert observer.

"Pennywhistle" groups comprised the only natives in Johannesburg who impressed me as acting spontaneously at all times. These were composed of ragged youths, each blowing off pitch on cheap slotted tin pipes, but in combination managing to produce a penetrating, syncopated type of music. There was something ominous about them, nevertheless, for they congregated arrogantly at the busiest street corners and were loath to move out of the way for white pedestrians. England's Prime Minister, Harold Macmillan, had not yet composed his telling phrase anent the "wind of change" about to sweep Africa, but those vagrant musicians were already tootling it for all to hear.

3

THE LAB

Luckily I did not have to drive downtown or through Joburg to get to the lab. From Rosebank it was a cross-country jaunt of some five or six miles. Within recent times the government had sponsored the preparation of live vaccines, for example yellow fever and poliomyelitis, and consequently had felt it expedient to create the Poliomyelitis Research Foundation for such somewhat hazardous work and to locate the unit in an isolated suburban area. The new laboratory built for those purposes was easily adapted to arbovirus studies, for many of the demands were identical—to name one, the maintenance and breeding of large numbers of experimental animals. (As a matter of fact, I was soon to witness a two-year battle among grown men for baby mice.)

At a distance the unit looked more like a small business plant than anything else, and the rows of assigned parking spaces fit the same scheme. But once within the doors, I could smell the mixture of chemicals and animal effluvia that pervades all biological laboratories. Open doorways revealed a few offices, but principally white-tiled hospital-like rooms crammed with work tables, sinks, centrifuges and deep-freeze cabinets. This was where serologists and virologists wielded their pipettes and syringes.

The Arbovirus Research Unit (ABVRU) had three entomologists on call, which would seem a more than adequate staff to maintain a flow of mosquitoes from bush to test tube. But Dr. Botha de Meillon, of the South African Institute for Medical Research, had all sorts of other commitments, both national and international. He pitched in whenever he could and had indeed been one of the key figures in the Simbu Pan expedition of 1955 when ABVRU made its initial discoveries of new viruses in Tongaland. James Muspratt, also of the Institute for Medical Research, likewise helped out then and at other times, though his personal investigative responsibilities often took him to distant places not connected with our program. Neither of these men could be put to the endless sorting and naming of collected mosquitoes unless each specimen was of sufficient museum value to be mounted permanently on a pin and given an everlasting label. That the insects should no sooner be identified than a laboratory worker crushed the juices out of them to inoculate mice with the resulting solution—such treatment immediately dissipated the aura of scientific entomology that field collecting had created.

Hugh E. Paterson, the third entomologist, undoubtedly must have responded in the same negative way to these dubious practices. On the other hand, he was still a student and had been definitely assigned to ABVRU, and therefore could be expected to obey the local rules whether he liked them or not. His absence was the result of youthfulness: the time had come for him to spend a final year of graduate study in England to acquire his doctorate in entomology, so off he went, ABVRU or no. Thus the mosquito net was readily relinquished to me by three far-from-reluctant abdicators.

My assigned desk turned out to be in Hugh Paterson's precincts —a spacious room on the second floor that could have served as a laboratory but was now lined with cabinets containing chiefly Hugh's collection of house flies. Until his return from England I would have this arena all to myself. When Ken Smithburn first showed me around, he pointed out one of the smaller cabinets, holding about a dozen shallow wooden drawers, which he said housed ABVRU's official mosquito collection.

"Hugh began it," Ken said, "and has been adding new specimens from time to time as something came along to fill a gap. Everything has been labeled, and many of the identifications were confirmed by Botha de Meillon and Jim Muspratt, especially in difficult cases."

Ken then picked up a volume from Hugh's desk, a copy of F. W. Edwards' *Mosquitoes of the Ethiopian Region. III. Culicine Adults and Pupae.*

"Be careful with this book," he cautioned. "It is out of print and this is the only copy in Johannesburg—borrowed from Botha's downtown headquarters, in fact. It will no doubt be your bible for the next few years."

With an already prepared and labeled collection of specimens in front of me, and a guidebook in my hand, I ought to learn to recognize African mosquitoes in short order. Rather, I *must* learn to do so, for that is what Ken expected of me at once. However, he seemed to recognize that his challenge might be a bit severe, for he promised to introduce me to Botha and Jim soon, so that I might ask them for help whenever I got stuck. But always he reminded me of the urgency of my task.

"We are planning our next trip to Ndumu, where the field station has been established, six weeks hence," he said. "By that time I want you to be competent enough to identify every mosquito that is brought to you by our native mosquito catchers. There is no use inoculating a mosquito into a mouse unless you know exactly what it is. Mosquitoes, as you know, have different biting habits, as well as other individual traits according to their species. Therefore, if you recover a virus from a mouse, you *must* be able to state the name of the mosquito from which it came. Otherwise you are left with no more than a virus, *period!* Well, you may know the date and place of the mosquito's capture—I'll admit that much—but if you want to go back in an attempt to discover how that mosquito became infected, acquaintance with its name will serve as a crucial guide to your investigations. Does it bite by day or night? Does it usually feed on birds or on mammals? Where are its resting places—in trees or at ground level? What kinds of water collections are utilized for egg laying and larval development? Could it be considered abundant or rare, and in either case, are its population peaks and declines related positively or negatively to seasonal variations in rainfall?"

I realized that those were only a few of the questions that an epidemiologist must ask of the entomologist, but Ken emphasized them to point out that the entomologist is taking great responsibility for a virus team. Every time he jots down his opinion of a mosquito's identity, everyone else follows it as gospel, because no one is there to contest it. Some virus outfits relegate the duty of identification to technicians. That is understandable, for as I have already noted, the sorting of mosquitoes all day long, day after day, is an onerous duty that first-class entomologists would rather avoid. They would prefer to be out making more stimulating investigations, and nobody can blame them. But here at ABVRU Ken had established the principle that no mosquito that had not been identified by a person who possessed a doctoral degree of any sort might be inoculated into a mouse. Oh, they accepted Hugh's diagnoses, but he was practically a doctor, which was why he had gone off to England now.

Ken's eyes were as penetrating as ever while he installed me in my lab and gave me his orienting lecture. His words both thrilled and terrified me. So I had six weeks in which to prepare, eh? Well, then, I'd better get on with it at once.

I took out the trays, one at a time, and carried them to my desk. The collection had been prepared meticulously. All labels were printed by hand in India ink, in tiny characters. The mosquitoes occupied a variety of arrested poses on their pins, death and desiccation having now rendered them so permanently relegated to those positions that the slightest touch or even a careless puff would knock off awkwardly extended legs, wings or antennae.

Before I came to Africa, someone in the Rockefeller Foundation offices in New York had provided me with a collection of reprints by Smithburn and his collaborators called "Studies on Arthropod-Borne Viruses of Tongaland." This I had probed assiduously.

Those seventeen virus isolations from Simbu Pan had almost all been derived from a mosquito named *Aedes* (*Banksinella*) *circumluteolus*. Obviously I must make the acquaintance of that insect first. It had been also the commonest species caught—no less than 4,657 of them! If it is humanly possible for entomologists correctly to identify that number of tiny two-winged creatures on the spot in the field in less than a month, using no more than a hand lens or low-powered binocular dissecting microscope for viewing them, they must be striking objects despite their small size. Moreover, they must look a lot different from their neighbors, if confusion could be so readily avoided. For remember: Botha, Hugh and Jim were all working under Ken's dictum: "If you call this beast *Aedes circumluteolus*, it had damned well better be *Aedes circumluteolus*. Otherwise, woe unto you!"

I had to search through the trays a long time before finding what I wanted. Heavens! How many kinds of mosquito was I up against? In India, where I had studied anophelines in connection with malaria investigations, I contended with only the single genus, *Anopheles*, and any fool entomologist can easily sort those out from the diverse array of other bewildering genera. If I were to study malaria at Ndumu, for instance, I would throw away everything else and concentrate easily on a mere twenty-four species of *Anopheles*. As it stood, my task was presently oriented to the culicine subdivision of the family (as shown in the list presented herewith), and after I had thrown away *Anopheles* I was confronted with no less than (as it eventually mounted up) 118 species distributed among ten genera, of which *Aedes* and *Culex* accounted for 51 and 37 species respectively, or three-quarters of the total fauna between them. What a morass this would be!

Six weeks? I sat back, turned my swivel chair, and looked out of the window in despair. Our wing of the lab overlooked a large sloping field devoted to raising crops of greens for experimental laboratory animals. In consequence it was constantly being fertilized and watered, and except for periods during the coldest time of winter, it was always lush. Now, in late August, spring was close at hand and the field had already turned green.

LIST OF GENERA OF MOSQUITOES OF SOUTHERN AFRICA

FAMILY CULICIDAE—MOSQUITOES

Subfamily	Anophelinae
Genus	*Anopheles*
Subfamily	Toxorhynchitinae
Genus	*Toxorhynchites*
Subfamily	Culicinae
Genus	*Harpagomyia* (= *Malaya*)
Genus	*Ficalbia*
Genus	*Mansonia* (= *Taeniorhynchus*)
Genus	*Uranotaenia*
Genus	*Hodgesia*
Genus	*Aedomyia*
Genus	*Eretmapodites*
Genus	*Aedes*
Genus	*Theobaldia* (= *Culiseta*)
Genus	*Culex*

Through the window I saw a Blackheaded Heron—much like our Great Blue in the States except for its terrestrial adaptations— stalk through the burgeoning vegetation and suddenly strike, not at a fish or a frog, but at a field rat. It tossed the animal up and caught it with its beak, dropped it, stabbed it, and tossed it up again and again until the rodent was insensible or dead. Then, with a final toss, the heron let the animal enter its gullet, and I watched a large swelling gradually descend through the bird's narrow neck. After that the feaster stood meditatively for a long time.

Meanwhile a great flock of Crowned Guinea Fowl had flown in from a grove of eucalyptus trees across the road. There must have been at least two hundred of them, but since I am a conservative in these matters, it may have been more like five hundred. If you have ever priced breast of guinea hen on high-class menus, you will realize that I was looking at several thousand dollars' worth of birds. Later I learned that Guinea Fowl are protected by law in South Africa, and that the law was rarely infringed upon by poachers because everyone is much more inspired to kill large quadrupeds. Until hunters have finished those off, and not until then, birds will be of only secondary interest to them.

The Guinea Fowl remained in a compact flock on the ground, gleaning whatever they had come for industriously as they actively moved forward. I reveled in the view of the field, the heron and its rat, and those clusters of speckled wild chicken relatives, for the vista from my latest office desk had been that of the Wistar Institute's incinerator chimney. Golly, and all other such expressions.

But this was treason. Why had I not been appointed ornithologist to ABVRU? Bruce McIntosh was the answer to that. My allegiance now belonged to mosquitoes. Reluctantly I turned away and opened Edwards' book. Not really reluctantly either, because I wanted to find out all about mosquitoes, too. There simply wasn't time to do everything at once.

Immediately, in the introductory pages, I came to a diagram that was to plague me for two years. I never mastered it, though I felt I ought to have. It displayed the lateral view of a mosquito's thorax, as if the insect had been stripped naked of all its scales but allowed to retain its endowment of bristles. Those setae occupied extremely critical positions on the cuticle, or exoskeleton, of the insect. These regions were marked off by lines or sutures into a dozen or more polygonal plaques, each of which had a name. The most important of them, apparently, was the mesepimeron. The mesepimeron is near the place where a mosquito's armpit would be if it had arms instead of wings, but beyond that I am unable to furnish a more accurate guide. Not only that (the location of that sclerite was hard enough, goodness knows), but I was obliged by Mr. Edwards to opine whether certain bristles were on the upper or lower part of the mesepimeron, or present or absent from either site. After that he said that I must count the setae, when present, as well as decide whether existing ones were stout or weak.

There was the matter of tarsal claws, too. And, now that I grimly think of it, pulvilli. Pulvilli are foot pads that enable many arthropods to walk on windowpanes, ceilings and other impossible surfaces by virtue of their being sticky or, occasionally, vacuum cups. Claws help these creatures hold onto uneven things like bark and wallpaper. My orders to examine those structures could not be followed. In the first place, I would have had to break off legs from the precious specimens in order to mount them on glass slides for study with a compound microscope, and in addition, the pulvilli, which are succulent in their expanded living state, were now shriveled beyond hope of determining what they had once looked like.

Indeed, I was stuck on the question of mesepimeral bristles for the same reasons. You simply can't take a mosquito off its pin without shattering it into dusty fragments. When an insect has dried with its wings covering the exact anatomical part at which you want to peer, you try to be philosophical but your blood pressure goes up anyhow. Besides, the bodies had become distorted during desiccation, rendering the thoraxes of most specimens— still fully scaled—so uneven that I often could not tell where the mesepimeron was even when wings were not in the way.

Nevertheless the collection was, as I have implied, eminently admirable. It was the book, then, that must next prove its worth. As in practically all entomological works, this one presented its clues to the identification of the different species of mosquito in a series of so-called keys. I can explain best what an entomological key is and how it works by inventing a situation that could help a Frenchman identify familiar garden vegetables by their English names.

Let us suppose that the Frenchman has an eggplant in his hand, having picked it from a grocery counter on which one can see potatoes, carrots, beets, radishes, onions, celery, corn, peas, lima beans and cucumbers besides. He does not know what to call his selection, so we slip him this little key which we happen to have in our pocket.

Key to Some Common Garden Vegetables

1. Edible portion rootlike 2
   Edible portion consisting of leaves, stems or fruit ... 5
2. Vegetable a tuber, without a leafy end ..... POTATO
   Vegetable a swollen root with leaves at one end 3
3. Vegetable yellow .... CARROT
   Vegetable red, white, or red and white 4
4. Vegetable red, eaten cooked .... BEET
   Vegetable red, white, or red and white, eaten raw ..
   RADISH
5. Vegetable consisting of edible stems 6
   Vegetable a fruit 7
6. Stems swollen, forming a bulb .... ONION
   Stems not swollen, elongated, parallel ...... CELERY
7. Fruit concealed by husks or a pod 8
   Fruit naked 10
8. Fruit within a cluster of leafy husks .... CORN
   Fruit within a pod 9
9. Pod cylindrical, fruit spherical .... PEA
   Pod and fruit flattened .... LIMA BEAN
10. Fruit globular, black or wine-colored .... EGGPLANT
    Fruit green, banana-shaped .... CUCUMBER

Couplet Number 1 promptly gives our Frenchman a choice of deciding whether the eggplant is a root or something that is *not* a root. He chooses the second alternative of the couplet—the non-root part—and finds that he is then directed to Couplet Number 5 to make his next choice. Here he obviously will select "fruit" instead of "stems" and thus he goes to Number 7. "Naked fruit" here takes him to Number 10, where he quickly decides that this is no cucumber but an eggplant. *Voilà!*

I found at once that the mosquito keys were a lot more complicated than that, but I had the advantage of those authenticated labels already proclaiming what each specimen was, so I simply cheated, working through the couplets in reverse until I learned whether a mosquito fell into a primary category with or without toothed tarsal claws or with pulvilli of some specified sort. Those damned bristles would fall into line by the same procedure—at least I would now know, when a specimen was supposed to have such setae, whether I could locate them or not.

Quickly I became lost in a revealing discovery. The introductory chapter had been far too frightening. Why, these mosquitoes showed up under a small magnifying lens as distinctly as the faces of movie stars! *Aedes circumluteolus* was a cinch, with its yellow-striped thorax. You couldn't miss the speckled wings and spotted legs of the two common large *Taeniorhynchus* species. *Culex tigripes* was unique in its genus. The three *Eretmapodites* all had silvery markings in different patterns.

Nevertheless, I found there were some species that resembled one another so closely that six weeks might still not be long enough to get them down pat. I took the labels off those and substituted numbers, for which I retained an index, so that as I ran specimens through the keys again and again (now in the forward direction), I would not inadvertently see the name tag as I was looking for banding or lack of it on leg segments. (Of course I replaced the labels eventually. To have removed them even temporarily was a scientific sin of grievous dimensions.)

After several weeks I could run most of the specimens down to a correct identification in jig time. But I became aware that I was doing well not entirely by use of Edwards' keys. I now remembered that the particular shards I was viewing had special qualities that had nothing to do with their classification. *Aedes cumminsi*, for example, had the left hind leg missing, and its abdomen was twisted to the right. *Culex univittatus* had lost its head. That is not the way to learn about wild living mosquitoes. At Ndumu I would have to examine not familiar pinned specimens, but buzzing creatures in glass tubes. My oh my! What would happen when I encountered those perfect ones? Would they be recognizable after this warped indoctrination?

Various other preparations had to be made. That was most

agreeable to me, because after I had looked through the dissecting 'scope for several hours, my eyes would begin to rebel. I would not be able to focus them on anything. We did have a coffee break each morning, which was a relief. And I could attend to a few necessary things at other times, when the print in Edwards' book could no longer be resolved into a legible form.

We needed cyanide bottles to kill mosquitoes, for example. After the insects' identification, they were to be "sacrificed" and stored in a dry-ice chest at once, in order to preserve such viruses as they might contain. Ken told me that the cyanide jars at Ndumu were undoubtedly old by this time, for no one had used them since Hugh's departure and they were then already becoming weak. Therefore I should make up some fresh ones.

"I don't know how to do it," I said. "I never prepared a cyanide bottle in my life, nor have I seen anybody make one."

I guess Ken thought I was a peculiar sort of entomologist, but probably he remembered that they had had to scrape the bottom of the barrel before I turned up, catching a dilettant instead of a professional, so he said nothing. However, he promised to arrange for me to be given the needed lesson. At the South African Institute for Medical Research (SAIMR) in Joburg was a very fine entomologist named Dr. F. Zumpt, who would surely cooperate. There Ken was overoptimistic. When he called Dr. Zumpt and explained our problem, that gentleman went into a rage. He said all the things that Ken had discreetly and politely kept to himself, namely, and in sum, that I must be a fraud.

Ken can be very persuasive. After a while Dr. Zumpt simmered down and said he would receive me, as a favor. Thus I came to meet a well-known authority on mites and parasitic insects. He had come originally from Germany but had now been in South Africa for a number of years. His Germanic background was evident in the neatness of his laboratory, the meticulous nature of his work, his no-nonsense conservation of words, and the impression he gave of being in a hurry to get on with things. My intrusion must surely be an annoyance.

Without wasting more than a minute or two on civilities, he opened a cabinet and took out a jar containing pellets of potassium cyanide that were kept under oil—perhaps it was kerosene—to keep them from absorbing moisture from the air and emitting deadly hydrogen cyanide fumes. Then he got a box of plaster of paris powder. With a pair of long forceps he removed a cyanide pellet—each was about the size of a pecan nut—and ground it under his heel on the floor. Then he swept the fragments onto a piece of paper and slid them into an empty mayonnaise jar. Quickly mixing the plaster of paris with water to make a thick paste, he poured the material onto the cyanide to form an overlay about one-third of an inch thick.

He made half a dozen bottles in short order. "Now we must let them sweat," he said. "You can't put mosquitoes in them and hope to get the insects out unless the jars are completely dry inside. Come back in three days and they will be ready."

He had a hot-air drying oven equipped with an exhaust fan in his lab. The jars were set inside, with their lids off, of course, and that was that. When the plaster of paris had dried, it would be porous, allowing cyanide vapors to diffuse slowly from below. I thanked Dr. Zumpt as adequately as I could in the few additional moments he realized were necessary to discharge minimum courtesy.

Actually I became quite friendly with that entomologist later, but not on the basis of cyanide jars. Now that I knew what to do, why should I bother him on that score again? However, when it came time to prepare fresh bottles, I found that we did not have the proper sort of drying oven at ABVRU. The exhaust must go up a flue to the roof, where no man, bird or beast can get near it. I did not have the face to presume on Dr. Zumpt's impatience or his equipment a second time. The solution? I switched to cigarette smoke for killing mosquitoes.

I did visit Botha de Meillon at SAIMR, as Ken had said I might, for help during my period of apprenticeship. A tall, thin man with wild hair, he was exceedingly affable; and he straightened me out, at least partially, on problems such as mesepimeral bristles. In most cases he could tell me about larger structures, more readily seen, that could be used to differentiate closely similar species. Without waiting for me to ask, he presented me with a copy of his book, *The Anophelini of the Ethiopian Geographical Region.* In addition he invited me to use the SAIMR collection whenever I wanted to. This was the official South African repository of insects of medical importance—not only mosquitoes but all sorts of other gnats, as well as tsetse flies and so on. Then he invited me to dinner at his home. In short, he became a great friend.

Botha (pronounced "Böta") introduced me to Jim Muspratt, a rather retiring, ascetic individual, who also pledged me the use of his collection, and then took me on to David H. S. Davis, currently the foremost mammalogist in South Africa. David's and my personalities embraced at once. He had come down from England as a graduate student twenty years ago to study plague and became so fascinated that he never went back to complete a few formalities that would have led to a doctoral degree. (That award has belatedly been accorded him in recognition of his work.) Botha (who grew up in South-West Africa) and David were devoted associates. At the time of which I speak they were just finishing the first volume of a treatise on South African fleas and their mammalian hosts.

Coffee breaks at the lab afforded me the chance to meet many other scientists in related fields of virology. The director of the Poliomyelitis Research Foundation, Dr. James Gear, rarely joined us. But the polio people, as well as groups working with Coxsackie and respiratory viruses, and collateral experts on things such as tissue cultures and electron microscopes, made those sessions educational as well as pleasant. Well, they were not always jovial, for when the conversation got around to South African politics, there was bound to be an acrimonious argument. Of course there was only one serious issue: apartheid. This is a topic on which you cannot take a mild stand. You are for it or against it, and let the blows fall where they may.

Whenever no one was unwise enough to bring up politics, the group usually ended up by discussing a phenomenon of which they seemed to be unanimously in favor. This was Dr. Robert H. Kokernot, a member of our ABVRU team who was currently on leave in the United States, having been engaged in local virus studies from the beginning as a sort of understudy to Ken Smithburn. Bob must have been a fabulous person to have so impressed everyone, and I was eager for his return. One day Dr. Gear came in, not for coffee but to show us a communication he had just received from Bob. This was a map of the United States, showing Texas as occupying about three-fourths of the entire area, all the other states being squeezed into the remaining one-fourth. Guess where Bob came from. Now I was not so sure that I did want to meet him. I could already hear him shouting "Yahoo!"

Doctor Gear's enthusiastic reception of Bob—the whole group's response to Texan manners—became more understandable to me after I had acquired a bit more experience of South Africa. That country is really something like Texas, with its vast openness and still pioneering spirit. Equate gold and diamonds with petroleum, and you achieve almost an economic parallel. James Gear would have liked to see a map of Africa with all the other countries crowded out, and I am certain that he could have learned to shout a falsetto "Yahoo" authentically. He invited me once to attend a party at his house on Guy Fawkes' Day, and I was petrified all evening while he and his brothers and their respective children set off fireworks—no, not as they were supposed to be detonated, but experimentally, to see what would happen if they tied two or three rockets together. Those missiles, consequently unbalanced, shot between our legs or landed on the roof of the house which (as I did not then realize) was made of corrugated iron sheets and could not burn. But *we* could have! This gaiety was quite in keeping with a Texan spirit.

Not that Dr. Gear and Bob Kokernot were other than exceptionally talented scientists, or insensitive to other human motivations. When they stopped playing like boys, they worked with the enthusiasm of top-notch intellectuals. Perhaps we less demonstrative, nonpioneering products of the Eastern seaboard are wishy-washy in both respects, though we are willing to grant that possibility only when it comes to shenanigans. On other occasions, when it was not Guy Fawkes' Day, I would be invited to meet persons such as Dr. Dart, the renowned South African anthropologist, who would lecture on his discoveries of early hominids and their primitive tools. Or perhaps it would be an adventurer sailing around the world in a small boat, or a *Time* magazine reporter stationed in Joburg. My irrepressible colleagues had spacious minds. If anyone had something new to say, they wanted to hear it, while I kept drifting off to some reflection about *Aedes circumluteolus* even while Dr. Dart was demonstrating how the hollow bone of an antelope's leg could be used as an apple corer.

We employed many other people at ABVRU—laboratory technicians, glassware washers, animal attendants for the mouse colony, and so forth, but I did not get to know any of them well because of my primary commitment to field work. Of course, I had to report my doings on paper from time to time, and our pretty secretary, Mrs. Grant, was a marvel in the way she could follow my scribbled notes.

That brings me to Pottie. Willem Hendrik Potgieter was a large man who served at almost anything we asked him to do—driving heavy vehicles and keeping them in repair, fetching the mail, going into Joburg to shop for this or that item, going to the airport to receive overseas shipments at odd hours—in short, he was the factotum that every well-organized laboratory needs. On a per man hour basis I spent more time with Pottie in South Africa than with any of my professional brethren, for he always accompanied me to Ndumu as aide, bodyguard, interpreter, companion and friend. As a result I came to know and admire him exceedingly well.

He was the descendant of an early Dutch settler of the same name. His father—an elderly man now, who must certainly have been involved in the Boer War—represented an outstanding exception to the commonly held notion that it is only the Dutch

Afrikaners who hold to the policy of apartheid. Indeed our professional-level kaffeeklatsch arguments at the lab were by no means lined up solely on the basis of protagonists' national derivations. Some people of British descent are today as vehement in defending apartheid as any of their Netherlandic compatriots.

Pottie's family had a farm in the province of Natal. I don't know how extensive or successful it was, but in the course of ordinary practice in South Africa the manual labor had been done principally by black natives who lived in their own kraals on the premises. Pottie's father, and *his* father before him, took an active hand in the work, but their responsibilities included a great deal of supervision and attention to marketing details as well, so that they could easily have kept themselves aloof and regarded the blacks as so many useful but insentient ants.

On the contrary, it had become a family tradition—to which Pottie adhered—to regard even the least of the Natal natives as deserving the regard of coequals. Obviously they were not slaves (though the system of apartheid as now practiced leaves the question open whether a benign type of slavery might not be preferable). In his boyhood, Pottie learned to take for granted that the farmhands would come to his father, singly or in groups, to ask advice or to seek his resolution of their disputes.

In the evenings little Pottie used to sit by the cooking pot in one or another of the kraals and listen to the old people tell their endless tales. He thus learned to understand and speak Zulu perfectly. His native tongue was Afrikaans, of course, and English should be called his *third* language as far as his proficiency was concerned, although he was good in that one, too, especially when it came to the profane vocabulary.

Nowadays Pottie's father was close to being an invalid as the result of several heart attacks. However, whenever he felt up to it and the weather was suitable, he would take off after breakfast to some rocky hills not very far away and sit all day among the baboons. He insisted that they would come to squat by him and to speak, and that he knew what they said.

Pottie had formerly worked as a policeman in Durban. I have

mentioned that he was large, but the description must be elaborated a bit further. I would hate to tangle with a 220-pound, almost-six-foot cop such as he, for although he looked very much on the fat side, an unusual amount of his underlying bulk was solid muscle. I commented on this one day at Ndumu when he had just challenged our fleetest mosquito catcher to a short footrace and had won.

"You should see my cousin, Doc," he panted. "He was chosen Mr. World for physical development a few years ago. He eats nothing but meat. His muscles are so big that he can hardly bend his elbows because his biceps hit his forearms when he tries to. He can't get a fork to his mouth. He is practically helpless. Why, shit, Doc, his wife has to shave and feed him!"

Pottie must have been the weakling of the family.

4

JOURNEY TO NDUMU

Ken interrupted my mosquito studies from time to time with further indoctrinating lectures. I soon learned that ABVRU's search for viruses in Tongaland had not begun quite so blindly as I may have implied. Before looking for viruses at all, Ken and his team had collected blood specimens from human beings and domestic animals in many parts of South Africa in a series of surveys and had tested them for immunity to the relatively small number of arboviruses known at that time. Antibodies, indicating immunity, were found in the blood specimens from Tongaland with higher frequency than in the specimens from elsewhere. Where immunity has developed on a wide scale, viruses must be present to account for that immunity, and so the chase naturally became centered in Tongaland.

Antibodies—chemicals elaborated by the body during the course of invasion and occupation by bacteria, viruses and other disease-producing organisms—may be life-saving if they destroy or immobilize the invaders fast enough. As long as antibodies persist in the bloodstream in sufficient concentration, they can prevent reinfection by the corresponding microbes, for each type of germ or pathogen stimulates the formation of its own specific kinds of antibody. Antibodies often provide information about the sorts of viruses present in a region and the kinds of host they infect. They may also reveal a pattern or periodicity in the times of descent of the viruses on various populations. Especially when long-lasting antibodies are involved—the types associated with lifetime immunity—one can quickly read the history of a virus in a village whose population is largely sedentary. If most of the adults give evidence of past infection but no child under, say, twelve, possesses antibodies, it is clear that the virus has been absent for over a decade. Or sometimes the pattern is occupational. Those persons who leave their village regularly to enter forests for whatever reasons will show immunity to a certain virus (having become infected in the forest by that virus), while other persons who rarely leave the same village will lack antibodies. Occasionally sera from domestic animals can be similarly informative after they are sorted into categories by age, sex or some other type of distribution.

From this approach to the study of arboviruses it is an easy further step to begin collecting serum specimens from various kinds of wildlife, especially birds and the smaller mammals. One now begins to learn the range of viruses not only in time and space but also within the framework of animal classification, and from here one can jump still further to a survey of biting virus vectors that infect those hosts, whether these feed indiscriminately on anything that moves or within more selective limits.

And as far as all that blood is concerned, why waste it all on antibody studies? In a few cases the doctor with his syringe might have appeared just when the human being or animal was experiencing the acute stage of infection (often without being ill). At such times the viruses might not yet have been overcome by antibody formation and would therefore still be circulating freely in the bloodstream. Thus any given blood specimen became a sort of grab bag: it might contain either a virus or antibodies. So why not use each specimen for a double purpose, inoculating a mouse with one fraction, to see if the animal would become virally infected, and saving the other portion for antibody screening? More than one virus would have been missed by ABVRU workers if they had not made their blood samples serve dual roles. Granted that the yield of viruses was much higher from mosquitoes than from serum specimens, Ken still would not have been satisfied to omit any part of the search.

Natives were almost invariably reluctant to being bled; in fact, they often refused. ABVRU consequently hit upon the euphemism "clinic" for bleeding sessions, and through Pottie and other interpreters bruited it about that bleeding was a form of "good medicine" benefiting one and all. Our doctors took temperatures at the same time, just in case a native happened to have a fever. Then, if his blood yielded a virus as well as assorted antibodies, we would have presumptive evidence that the virus had produced a febrile response in the ambulatory patient. Admittedly it was not a very potent virus.

The original Simbu Pan expedition discovered three new viruses, which became known as Simbu, Spondweni and Pongola. But in addition the laboratory had detected strains of three already known virus species in the mosquito suspensions. Each one made history by showing up in Tongaland, for the viruses had diverse backgrounds far removed from the lowlands of Natal province. Rift Valley Fever virus was first isolated in Kenya in 1930 during an outbreak of disease in sheep. Wesselsbron virus was found in Orange Free State province in South Africa in 1955 under remarkably similar circumstances, ewes aborting and young lambs suffering a high mortality just as in Kenya. Bunyamwera virus, isolated in Uganda in 1943, was a Smithburn discovery from the yellow fever era; it came from a monkey that died after the routine inoculation of wild-caught mosquitoes.

Thus the Tongaland arbovirus list stood at six from the word go. Perhaps there were no others. Even if I identified mosquitoes "until my eyes dropped out," the lab in Joburg might simply recover more and more of the same. That would soon become a sterile performance. To avert such an outcome, I determined to collect mosquitoes on as broad a front as possible, and would try to detect the best sources of the uncommon species and intensify the search in off seasons such as the hotter dry months when mosquito-breeding waters dried up. That would keep me all the happier from a mosquito standpoint for I would be playing with the widest possible spectrum of species, but I must always remain aware of viruses in the background: without those submicroscopic props, my game would lose its sponsors and I might be back at public health work in Philadelphia, regaled only by the natural history of runny noses and sore throats.

The six weeks of my apprenticeship took a long time in passing. Impatient as I was to try out my new knowledge at the field station, I simultaneously became more apprehensive as each day went by, feeling that I had not learned enough and that my pose as an instant expert was becoming progressively more ridiculous. Meanwhile I attained the age of half a century and became a grandfather, though those seemed minor triumphs beside the mastery of mesepimeral bristles. But now—at last—it was time to go.

ABVRU's field van was a Land Rover that had been converted by our lab carpenters into a small bus of sorts, with room for three to sit in front and with folding side benches in the rear that could accommodate at least a dozen mosquito catchers. Those seats were now hooked against the walls to make room for all our equipment. We had loaded most of the things yesterday afternoon. My chief concern had been to see that no entomological need was overlooked: cotton-plugged glass tubes in abundance, of course, but especially the new cyanide jars and Edwards' book. I must have looked a dozen times to be sure that Pottie had not left them out as he rearranged articles in the annoying (and, it seems to me, unnecessary) way that packers invariably do.

There was no question of taking my Renault or Bruce's car to Ndumu. Perhaps they could have made it without falling apart, but that would have been only good fortune. Even the Land Rover was known to have become stuck, requiring oxen or a tractor to pull it out of ditches. This trip was no mild outing. From

Johannesburg to Ndumu one had to cover about three hundred fifty miles of highway, commencing with a good paved surface and progressing to worse and worse roads until they scarcely deserved to be called more than cleared areas between flanking bush.

I had obtained a couple of bananas from the Garden Inn dining room last night at dinner, and therefore put something into my stomach while dressing. Now, at three o'clock in the morning, Bruce, Pottie and I converged at the lab, and as we finished loading the van, that meal did not feel very substantial. An African night guard lowered the iron gate after we drove out, and at last we were on our way—but whither? Our preparations remained incomplete. First we must make a stop at the chemical plant in Germiston to pick up a supply of dry ice. Among other paraphernalia that Pottie had stowed and restowed so painstakingly were two insulated canvas containers that would each hold four foot-square blocks of frozen CO_2. Without this material we should never be able to get virally infected mosquitoes back to Joburg, and one might as well give up the expedition. The canvas receptacles were now stationed at the very rear of the van to make them immediately accessible from the door.

Therefore Germiston was a key point on our trip, for all its gloom and grime as a dreary mining and industrial town. Before daybreak it was always cold there, whatever the season. Acrid smells hung on the predawn air. Batteries of klieg lights lit the heavily guarded entrance of the chemical plant, as well as its many tall fractionating columns and chimneys needed for making sulphuric acid and other molecular workhorses.

That first time, I had no idea how long two bananas must sustain me before I could enjoy (hardly an adequate verb) a good breakfast. We had worked hard loading the van and now we labored to get the dry ice blocks into their insulated canvas conveyors. At last Pottie took the wheel and effected our exit from the chemical plant, but not before we underwent a final security checkout at the entrance office. Then the gates opened just far enough to let the Land Rover pass before they closed emphatically behind us.

Now for the long run. Daylight was appearing, and Pottie had no thoughts of sparing the vehicle its maximum capabilities. He must have been hungry, too—I think he was always ready to eat. We sped—which was possible on this first portion of the journey —for about two or three hours, during which I became more and more faint, until we reached Ermelo, a town in the very midst of the Highveld—a vast irregular plateau occupying much of South Africa's interior—that supported a Greek restaurant. "Greek" was not important at that moment, but it so happens that Greeks are the chief restaurateurs in South Africa, just as they are elsewhere. I would have welcomed any other nationality, as long as I could build on the wreckage—nay, extinction—of those bananas.

Now I was about to learn one of Pottie's traits that really endeared him to me. You could always eat heartily in his presence, because his enthusiasm was like a shot in the appetite center, wherever that lies in the brain, and provided you needed stimulation. He and Bruce both recommended eggs, steak, potatoes, tomatoes, bread and coffee, to which I acquiesced except for the eggs. (What of my insipid tomato juice and corn flakes? Could I face up to anything as virile as steak for breakfast, no matter how close I was to syncope?)

After what seemed a long time, we were served with huge platters of food, steaming hot, and Pottie went into the act I wish to describe. He didn't say grace, certainly, but the equivalent was there, and his Maker must have been gratified when Pottie's message reached him. Spreading his arms apart, resting his hands on either side of the plate, and bending down to sniff aromas rising from the food, Pottie proclaimed, "Ah, this looks *good!*"

Whereupon and forthwith, keeping his elbows still outspread, he attacked everything at once. Fat, gristle—parts that I carefully trimmed away—went down his gullet indiscriminately. To him they were as excellent as the meat. Bruce did almost as well, I should add. As for me, I probably ate the largest breakfast of my life in Ermelo, and a lucky thing I did so, for we had not yet come to the rugged part of our trip.

The road thus far had been hard-surfaced, allowing us to make good time. This area was almost devoid of trees; even such as there were had been hand planted. The terrain was divided between grazing lands and huge fields of maize or, as Afrikaners and natives alike call it, "mealies." Both sides of the road were fenced, to keep cattle off the highway. Parallel embankments, thus protected from browsers, supported an endless show of wild cosmos in pink and white. I think they were not really wild but had established themselves as escapees from someone's flower garden, for they had not reverted to a dwarf form, nor was there anything else inferior about them.

But now the pavement gave out and the dust and bumps began. At the present season, spring-summer showers were just commencing, so that one stretch of road would be parched, another muddy, and still another almost washed out by a local deluge. There was nothing boring about that drive, especially while Pottie was at the wheel, for he seemed to think that he could overcome deficiencies of the road by an act of will, transmitted somehow to the Land Rover, so that it would skim over breaches as if they did not exist. From time to time Bruce would arrive at the end-point of his ability to bear up under this tension. "I'll drive now for a while," he would say, and then for a time I was able to relax, too.

Having been brought up in Natal province, Bruce shared Pottie's enthusiasm for the region. Indeed, he too was a third-generation Afrikaner, although his medium stature, broad shoulders and heavy dark eyebrows proclaimed him immediately a Scotsman. Bruce likewise had learned to speak Zulu from the natives as a boy and was thus able to get things done in the bush with the most enviable ease.

After we crossed the Assegai River and continued east, towns became more sparsely distributed. The last one on the Highveld, Piet Retief, was fairly large and really the most attractive town along the entire way. Then, not many miles beyond, we suddenly reached the jumping-off place. I was glad that Bruce held the wheel as we descended a winding road into the lowlands. Everything changed at once. The landscape was a great hodgepodge of tumbled rocks and irregular bare pinnacles in total disorder. Brushy tracts, interspersed now with small but truly indigenous trees, bore no resemblance to the Highveld we had just abandoned. Candelabra-like euphorbias bespoke desert conditions.

The road led up and down, after each rise attaining a lower level, until we descended a final stretch into the small village of Pongola. A fairly sizeable and constant river of the same name here provided enough water to support an irrigation scheme for the cultivation of sugar cane. After lunch at an Afrikaner restaurant, we paralleled the Pongola River for a few miles, having first crossed it and finding ourselves flanked by cane fields on the left. The river then veered southward and we continued over several successive hills until we arrived at a tricornered geographical junction: the end of Transvaal province (which had been our venue all along), the northwestern beginnings of Natal province, and the southern extremity of Swaziland.

At this tricorner it was tantalizing to enter Natal but to have to leave the province at once in order to get across the Lebombo Mountains before dropping down finally to the coastal plain. To do that we had to cross a tip of Swaziland. Customs practices seemed *not* to be practiced at this boundary. There was an official sign and a gate, but no one attended to the rare passage of motor vehicles. Bruce suggested that it might be refreshing to stop at the broken-down hotel in Gollel for a cold beer, and to that I agreed, though for once I would rather have kept going. Fatigue was upon me in only a slighter degree than my eagerness to see Ndumu.

Following the intake of *two* beers in Gollel, we made a left turn to the north and stopped at a "petrol" station to fill our tank. In this British protectorate the natives were allowed to drink, and I was impressed to see a drunken black across the road displaying his large but limp penis before a drunken female in broad daylight. At least they could dream openly and contentedly in Swaziland.

The Lebombo Mountains took us again into uplands, but not before we had passed through some gorgeous lowland scenery, abounding with rushing streams. At one point Pottie (again at the wheel) stopped under the shade of a grove of fever trees so that we could relieve ourselves of the beer. Fever trees are a species of acacia, but much more ambitious than their more ordinary congeners. They rise to great heights, with gracefully spreading limbs, and have a soft bark of corky texture. Their name derives from an exorbitant need they display for water. If you see fever trees, you can expect to find water, and, as an obvious corollary, malaria— an association that was sensed by human indigenes long before the aquatic phase of the malaria cycle had been indubitably established.

Now we were on the last lap, and Bruce, knowing what would come next, firmly took command of the driving. After we passed through the mountain village of Ingwavuma, a terrifying road led over several craggy shoulders and then down to what was essentially sea level. At this point we had left civilization behind. Nor was there any longer a true road. Wherever traveled ruts became too deep, people simply hacked a detour around the impasse through adjacent thickets. So many choices existed along the way, as a result of that practice, that we often could not tell which of several forks to take, although only one of them was the currently passable route.

Our speed could no longer be discerned on the speedometer. Those last miles were agonizing. Eventually we came to an open area devoted to tobacco farming and the cultivation of sisal for rope making, and immediately after that we reached the only country store in the region, managed by Mr. Duplessis, or "Dupe," as he was called. Bruce said encouragingly that we were now only six miles from camp.

The afternoon threatened to give way to evening. Since we had been on the go for more than twelve hours, I scarcely cared what happened next or what our destination would be like. Six miles: then I could get out of the vehicle and lie down, if only on the bare earth. We passed a gate with a posted sign saying "Ndumu Game Reserve," but a wandering hippopotamus would hardly have aroused me at that point. And thus, finally, Bruce brought the van to a halt in front of a low building with a corrugated iron roof, in the midst of a large bare fenced compound. A tank for collecting rainwater from sloping metal sheets, as well as an outlying sheltered latrine, testified to the forethought that had gone into construction of this elegantly appointed haven.

5

TONGALAND MOSQUITOES

My indoctrination to new places seems often to depend on the voices of pigeons or doves. On our first late afternoon at Ndumu I heard a doleful note that Bruce said came from an Emerald-spotted Wood Dove. Eventually I came to know that common bird intimately, and I still become nostalgic when I recall its cooing. We have our Mourning Doves in the United States, and they *do* mourn, but this species—a small brownish bird with a number of metallic green splashes on its wings—had a longer lament to utter. Roberts, in my ornithological bible, paraphrased it as a series of "du du's." He went on to say that various native tribes translated it as "My mother is dead! My father is dead! All my relations are dead! Oh, oh, oh, oh, oh. . . ." However, Pottie improved on that version by giving me both the Zulu impressions of the sounds and their English equivalents, as follows:

> *Ndon, ndon, ndon.*
> *Ngange né ngange zafa;*
> *Ngange né ngange bazeta;*
> *Man je inholizian yame*
>   *ilokee yahlala iti*
> *Ndon, ndon, ndon, do, do,*
> *Ndon, ndon.*

Dōn, dōn, dōn.
I had babies and they died;
I had babies, they were stolen;
And now my heart
    is going
Dōn, dōn, dōn, dō, dō,
Dōn, dōn.

Our "home" was known as the NRC camp, which sounds like the National Research Council but stands actually for Native Recruitment Center. I was introduced at once to Jack Tschembene, our resident cook and housekeeper, who had been "recruited" twice himself—once for the ordinary function of working in the mines at Joburg, and the second time for another ordinary cause after he had murdered a man whom he surprised in cohabitation with one of his wives. He served two years in each instance, the second time at hard labor without pay. His broad smile now accorded with our distance from Johannesburg.

Jack might have been a model for Madame Tussaud's museum; he was a Zulu prototype. He was not handsome in any sense, but his proportions and ruggedness were unequaled among other Zulus I have met. Even the veins of his forearms and arms stood out as if they had been carved in ebony over muscles that bound them without need for skin to hold them in place. The closest things I have seen to such perfectly executed sculptural relief are the diaphanous folds of the robe that enwraps the thighs of the Victory of Samothrace in the Louvre. Jack had those graven attributes, but no one came to Ndumu to put them down in stone.

We were still unloading the van when the first mosquito catchers arrived, not to begin collecting insects but to help with whatever chores Pottie might give them. In short order they were carrying things into the house, helping to unload dry-ice blocks and to get them into the large insulated wooden chest that had been made for them, and then, at Pottie's command, they ran out and formed a bucket brigade to fill the steel drum on the roof that served as the source of water for our indoor shower. One boy manned the pump in the yard, others carried buckets, two or three stood on a ladder to pass the buckets upward, and the last one dumped them into the barrel.

I was instantly amazed to see the good fellowship among these youths, who ranged in age from about ten to eighteen, and their free approach to all of us "white folks." I would have given a lot to be able to speak with them, but they knew no more of English than I of Zulu, and I had to rely on interpreters—mostly Pottie, but also Bruce when he was there.

The boys varied widely in appearance. Tall, fine-looking ones were said to resemble Zulu ancestors, while shorter, more plain-featured individuals stemmed presumably from Tongas. Tongaland, as a real political region, does not actually exist, but this portion of northeast Natal was formerly occupied by the Tonga tribe before it was overrun by Zulu warriors. The women were spared and some still speak Tonga, but all males scorn that tongue.

Pottie took pains at once to acquaint me with a slender small boy with rather light skin who, he said, was one of the most intelligent and would act as my entomological assistant, rather than as mosquito catcher. This was Qmba Ngwenya. That name being a bit difficult, Pottie had dubbed him "Hut-nut," a sort of affectionate diminutive for Hottentot, in reference to his light coloration. (I had not realized that Hottentots were not fully black.) It was not unusual to find light-skinned Africans around Ndumu: that was said to be due to the proximity of the Portuguese province of Moçambique, just across the Usutu River. Racial barriers were not officially recognized there, and some Portuguese blood trickled into this part of northern Natal from time to time.

I asked Pottie whether "Qmba Ngwenya" meant anything in particular. "Oh, yes," he answered. "It means 'blown-up crocodile' —you know, when it has been dead and lying in the sun for several days."

"Why on earth should anyone name a child that?" I asked.

"You see, Doc," said Pottie, "these people have their own kind of humor. They often choose names that are the craziest they can think of. That little mosquito catcher over there is named Magwalo, which means 'rubbish.' But they sometimes make up names that don't mean anything at all. They like those for the sounds. Another of your mosquito boys is named Dom-Dom, and Dr. McIntosh has a bird catcher, who is not here now, called Po-Po-Po."

Dom-Dom (pronounced dum-dum) was of the Zulu type, an adolescent just stringing out into manhood. He had large eyes and a delicate, almost feminine softness about him which again was totally in discord with my preconceived ideas of savagery. His older brother, Mkawpis, was the more frightening type that Dom-Dom would probably become, but even he, in his loincloth decorated with scraps of fur, was genial and filled with no more than healthy mischief. These two were among our best workers. In defense of the rest, I must say that none of them really shirked—it was only that some were more able than others.

The cots having been set up and the tank filled, Pottie told the boys to go and to return in the morning for a day of collecting. Now I had a chance to look around the house, carrying a glass of Medical Reserve (an excellent South African brandy) mixed with water from room to room. The front of the building was occupied by a long screened porch that had been converted into two laboratories, entomology on one side and a larger all-purpose area on the other. Behind that were two inside rooms, for dining and for storage respectively. A small wing off the dining room was the kitchen, while the storeroom led to another screened porch at the rear that served as a dormitory. A door from the latter led to our fabulous shower. Jack Tschembene lived in a small, windowless, one-room structure some twenty paces distant.

That dormitory-veranda looked good to me at once, and not only because I could have used it immediately without troubling with dinner. Screened along its entire front from floor to ceiling, it gave one the impression of being almost out-of-doors, which is of course the best possible place to sleep. I did eat dinner, but then yielded to the invitation and sank into my cot. Bruce and Pottie turned out to be great snorers, but they did not get together in either their rhythms or harmonies. Jack must have been an insomniac, for I heard him stirring in the kitchen while it was still dark. However, I soon blessed him, for with the first appearance of light he loomed with *bed tea,* a cult or addiction or vice—whatever it should be called—that I had embraced in India but had now been deprived of these many years.

What pleasure it was to lie in bed, smoking cigarettes, drinking tea and talking with Bruce and Pottie about all kinds of inconsequential things! Birds were singing or calling outside, and though I hadn't the faintest idea what they were, I realized that eventually I should find out their names, and meanwhile this idyll was too delicious to violate with ornithological questions. Then suddenly I thought: Mosquitoes! This was the day on which I would be put on trial. I imagine a condemned murderer sometimes awakens and does not realize immediately that something particular is going to happen to him on that morning.

Jack prepared sausage and eggs for the others. I had made it plain that I must have bacon, so he gave me that with a large steak thrown in as an afterthought. The mosquito boys showed up, and I distributed three boxes, each containing about a hundred cotton-plugged glass tubes, among them. Pottie told them where to collect, and they trotted off, arguing among themselves about who would have the privilege of carrying a box on his head.

It would then take them most of the morning to fill the tubes, even though only a single mosquito must go into one container. They had been trained by Hugh and Botha to recognize not only a mosquito as such, but *female* mosquitoes (which are the virus carriers) and also female *culicine* mosquitoes (which have short palpi compared with female anophelines). They could make that discrimination with the naked eye! Their methods included several techniques. In seasons when mosquitoes were abundant, they would simply sit down in a thicket and catch biters that settled to feed on them. Or, working in pairs, they collected alighting insects from each other. At less propitious times they would walk through grass and weeds, stirring up resting mosquitoes, and then watch them in flight until they landed. That, to me, was the most outstanding feat of all, for though I tried to do it many times, I invariably found that a mosquito that came to a perch became invisible—it out-and-out disappeared.

Whatever the position of the mosquito—on the skin, or on a twig, leaf or grass stem—the catcher now removed the cotton plug from a small test tube and crept up on his quarry. A deft motion brought the plug back in such a way as to drive the mosquito into the tube, unless it escaped beforehand. Naturally there were many misses. Each catcher kept his own score, and rivalry among them contributed highly to ABVRU's welfare. Dom-Dom's soft eyes and Mkawpis' mischievous ones were sharp enough to keep the brothers at the top of the list.

"You'll have to wait quite a while before the first box comes in," said Bruce. "Let's get into the Land Rover and I'll show you all the collecting stations, as well as some of the other local scenery. We'll see some birds, too. After a couple of hours we'll come back, and Jack can give us mid-morning coffee." Further luxury!

We drove first to Site 18. This was where Pottie had told the boys to go, and we found them stalking insects like small Sherlock Holmeses. At least their bent-over postures needed only the addition of a magnifying glass to complete the picture. They were unable to see undisturbed resting mosquitoes. Their ruse, therefore, was to walk along slowly, shuffling their bare feet through grasses and fallen dead leaves. Once a mosquito was nudged into flight, the boys were on its trail. I looked into one of the boxes and saw that it was already half filled with occupied tubes.

My entomological predecessors at Ndumu had chosen a variety of mosquito-collecting stations more or less at random to find places where mosquitoes could be obtained most abundantly, or whence viruses might be derived. About half a dozen of the best stations were still in use. The established routine was to work them in rotation, because daily raids on the same place did not allow enough time for the mosquito population to replenish itself.

Site 18 became one of my favorite places to watch birds or even just to walk. It was a discrete feature in the local landscape, being a sort of island of giant fig trees in the midst of open cultivated land. The crowns of these figs formed a continuous canopy so that it was relatively cool and crepuscular at ground level even on hot days. That was one reason why the grove had been selected as a collecting site, for mosquitoes are delicate creatures that cannot stand excessive heat or direct sunlight, both of which dry them up in short order. Hence insects of this class that flew about freely at night, biting natives and their animals as well as any wildlife that roamed in the surrounding flatlands, might well seek the fig gallery at daybreak in order to find shelter during forthcoming torrid hours. That reasoning proved to be correct. While mosquito catchers could fill tubes rapidly in the grove, they were hard put to it to make a single capture in adjacent weedy fields.

Another motive for the choice of Site 18 was that the virus team wanted to collect mosquitoes from the canopy, and not many canopies existed hereabouts. One did not know where viruses might come from, except that no possibility should be excluded until it had been thoroughly tested. Yellow fever virus, for example, was known to be transmitted by canopy-inhabiting mosquitoes to equally arboreal monkeys in the tropics of both Africa and South America, though Ken's serological surveys in South Africa had not disclosed yellow fever in natives living that far from the equator. Yet there was always a chance that it had been missed. The coastal plain of Natal lay in what has been termed a "tropical corridor" by biologists in several different specialties, because various groups of animals and plants, characteristic of Central Africa, have been able to invade this southerly coastal region while failing to colonize the more temperate Highveld.

Moreover, aboreal primates were known to be present in and about Ndumu. David Davis had collected three species at Simbu Pan: greater bush babies (or Galagos), vervet (or green) monkeys, and samango (or blue) monkeys.

Hence it was up to mosquitoes to do the rest in order to complete a treetop yellow fever cycle. Our native helpers had been instructed how to nail a series of cross-pieces up the trunk of one of the tallest fig trees and then prepare a flat platform on which they could sit while catching mosquitoes forty-six feet above the ground. I never ventured to climb to that aerie, though I am sure that Hugh Paterson and Bob Kokernot did many times. The insect catchers could do it as if they were four-handed primates themselves.

I have described Site 18 fairly fully in the positive sense but have not presented the facts that it was *not* adjacent to a river or pan (as local ponds were called), and that it was *not* close to known breeding places of mosquitoes. Other sites had one or another of those qualities, or a combination of them, and were either occupied by trees or only thorn scrub. Some were transitional in many of these respects, consisting of older scrub growing into forest. It is probably obvious that collecting in varied habitats will yield more kinds of mosquitoes than if the search is confined to only one. In addition, it happens that even when mosquitoes of the same species are found in two different kinds of habitat, their numbers almost invariably show a significant excess in one area compared with the other. That eventually gave us the ability to spot locations in which it would probably be possible to obtain a given kind of mosquito in useful numbers.

We needed such precise information in order to narrow our search for the source of viruses. Whenever the lab detected sick baby mice following the inoculation of mosquito suspensions, they immediately telegraphed us the news through Dupe's store six miles away, specifying the kind of mosquito concerned and the date and location of the capture. At least ten days would already have elapsed since the original mosquitoes were caught, but we nevertheless revised our schedules immediately to put on a drive at the indicated collecting site not only for further mosquitoes but also for blood samples from small mammals and birds.

Bruce showed me several other collecting areas, both current and abandoned, until thirst lured us back to camp for the promised coffee which was—and what else could it have been?—Nescafé, served by Jack with Pongola sugar and evaporated milk.

Then the first group of boys, headed by Dom-Dom, arrived with a box of filled tubes. I carried it into the screened entomology cubicle to see what I could do. Those wretched boys clustered about outside, with their faces pressed against the screening, to see the new doctor at work! Not that they would know whether I was doing right or wrong, but I was a novelty in their midst, and they had no inhibitions concerning impolite staring. I decided right then to adopt a completely rigid and dead-pan form of behavior, whether I was stuck on a specimen or knew it at first glance. Soon they would tire of watching if the caged entomologist refused to exult or growl. They eventually gave me a name in Zulu which Pottie translated as "The Quiet One." They had titles for everyone else on our team, more animated but less flattering, so I'll not try to remember them.

We had brought along a binocular dissecting microscope, much to its disadvantage in the rocking van, but I wanted to make my first examinations more simply, with a jeweler's 5X lens affixed to a spring-steel loop that extended to the back of my head. Pottie helpfully explained that Botha and Hugh used to place the tubes, once the contents had been recognized, in shallow white enameled pans on which they could write the species' names in wax pencil. Thus they gradually accumulated designated lots of *A. circumluteolus*, *T. africanus*, *C. univittatus* and so on. Each lot was then counted and passed on to Qmba Ngwenya for cyaniding. After a day's work the names could easily be rubbed off the enameled surface with a cloth, and the trays were ready for further use.

So what else did I need? The Edwards book, of course. And—courage. I affixed the lens to my eye, picked up the first tube, and squinted. The glass was practically opaque! The mosquito catchers had no idea of cleanliness. As they went about their forays, they carried five or six tubes in one sweating hand, returning to deposit them in the box only when all were filled. The moisture

caused soil and leaf mold to adhere to the glass. But this obfusca-
tion had a more glutinous basis than mere sweat and mild organic
rot. Practically all the boys—especially Magwalo—had constantly
running noses, and since they had never heard of handkerchiefs,
they wiped away secretions in the time-honored way of the poor
man who discards what the rich man puts into his pocket.

I called Pottie and asked him to tell the boys to wipe the tubes
so that I could see what was inside—and, moreover, to do that in
future every time they returned from the field. Maybe my order
was what they had been waiting for, because although they
thought it was funny that the "new" man should at once complain,
their response was a bit sheepish. Perhaps I was a "Quiet One,"
but I was not going to stand for any slipshod nonsense. That message remained with them as long as Pottie was there to back me
up. During his absences, when other problems arose, I had to be
my own factotum, for even when I demonstrated something I
wanted done, the boys looked at me and at each other absolutely
blankly, as if I were some sort of lunatic. That might have been
willful, though I imagine they had so little understanding of what
we were trying to do that even their own performance—the not
ordinarily possessed ability to recognize female culicine mosquitoes from among the rest of a vast insect fauna—was not at all to
be recognized as an intellectual accomplishment but simply as an
automatic act at the moronic level.

After half an hour the tubes were cleaned and I began again.
Immediately I thought I was in further trouble, for the first mosquito I tried to study was buzzing around in the tube making
nothing more than a blur of itself. When it did finally settle, it
rested on a curved part of the glass that distorted its shape by
refraction, and when I turned the tube to get a direct view of the
insect, it began to fly again. However, after a few moments it
came to rest once more, this time exactly opposite my lens, and
glory be! there was *Aedes* (*Banksinella*) *circumluteolus* just as
plain as day. Indeed, I soon learned that those painstaking studies
of dried, faded and grotesquely posed pinned specimens in Jo-

burg had been more than adequate exercises for the job I now had to do, because these freshly caught living specimens were much more brightly colored, and their stance when resting with wings folded over the back made all parts of them except the upper surface of the abdomen easy to see. Of course I ran into "elderly" mosquitoes that had lost some of their scales, and the mosquito boys damaged quite a few while making captures. But there were some fresh ones in every lot, and once I became familiar with those, it was only rarely impossible to recognize less perfect ones.

I called Qmba Ngwenya to begin killing. We kept our cyanide jars tightly capped when they were not in use, but now I unscrewed the lids and covered the mouths of the bottles with sheets of thin rubber, in the center of each of which a slit about three-quarters of an inch long had been cut. The sheet was stretched taut, so that edges of the slit remained touching, and a rubber band around the neck of the jar held the lips in that position. As Qmba Ngwenya removed the cotton plug from a tube, he quickly popped the open end through the slit, and the contained mosquito either flew at once into the cyanide fumes or else had to be induced to do so by his tapping the tube and shaking it up and down. Some mosquitoes escaped at the instant the cotton plugs were withdrawn. Then we would chase them around the screened insectary until we managed to recapture them with glass aspirator tubes.

After a species "lot" had been consigned to the gas chamber, one mosquito at a time, Qmba Ngwenya removed the rubber sheet and inverted the bottle over a funnel leading to a small Pyrex tube (actually a "Wassermann" tube) to shake the mosquito into this second container. He then inserted a paraffin-coated cork and pressed it firmly into the tube. He put the tube in the now empty enameled pan that still had the name of the species written on it in wax pencil. I now put the name on a narrow strip of adhesive tape and wrapped this around the tube. The lot number was noted also, so that it could be looked up in a duplicate notebook that I kept, listing under each number the date, collecting site, species, and number of specimens in each tube. One copy remained with me at Ndumu while the original accompanied the tubes when they were shipped to the lab. My last act was to get each tube into the dry-ice chest as quickly as possible. The reason for paraffin-coating the corks was to prevent CO_2 from diffusing into the tubes. Even at $-60°C$ viruses survive, but they could be inactivated by the presence of that gas, because it apparently dissolves in water and forms carbonic acid which, although weak, is sufficiently noxious to delicate viruses to destroy them.

The broad yellow dorsolateral thoracic stripes of A. *circumluteolus* were like banners proclaiming its identity. I had to glance at the wings in each instance, for another member of the *Banksinella* group of *Aedes* was found occasionally at Ndumu. This one, A. (B.) *lineatopennis,* had a longer streak of yellow near the forward edge of the wing than did A. *circumluteolus.* However, that seldom gave much trouble except for extremely ragged specimens. One might think that these could simply be discarded. If we had been collecting mosquitoes merely to assemble an entomological display, that is undoubtedly what I would have done. On the contrary, in a virological program, old specimens might be the most valuable of all, for these would have had time to take more blood meals than their younger sisters and therefore the chances of their having acquired virus infections were correspondingly heightened. Consequently I spent a heartbreaking and eye-wearying amount of time studying derelicts, in the end discarding only a few of them. For of course I *had* to throw those out according to Smithburn's Law: No unidentified mosquito shall be inoculated into a mouse in my laboratory!

Among the tens of thousands of the *Banksinella* category that passed under my eye—I think it was around seventy or eighty thousand—no more than a few dozens were A. *lineatopennis,* the rest belonging to the common species. Again, you might think it hardly worthwhile to have had to look at seventy or eighty thousand pairs of wings of *A. circumluteolus* merely to sort out those infrequent congeners. But if a virus were ever recovered from *A. lineatopennis,* one would have been set on a new trail of investigations to look into its distribution, breeding places, blood-feeding choices and so on, in case those attributes should prove to differ from those of *A. circumluteolus.* Every one of those multitudinous *Banksinella* monsters required that I be as keen-eyed and as much on the alert as if it were unique.

The first five or six mosquitoes I had looked at were all *A. circumluteolus,* but then I came to a somewhat less sturdy specimen. It was slightly smaller as well, and the lack of thoracic banners or bright markings of any kind brought me back to a state of near panic. But wait! As I slowly rotated the tube, I glimpsed a pale stripe on the anterior surface of the hind tibia. Aha! Doctor *Culex univittatus,* I presume. And so it continued as I met additional species, until the box was empty. The other two boxes arrived in due course. My enameled trays held over a dozen categories of mosquito: great heaps of tubes containing *A. circumluteolus,* a respectable number with *C. univittatus,* somewhat less respectable tallies of the two species of *Taeniorhynchus,* and a smattering of other kinds of creature, in some instances there being only one representative.

Yet this had been almost a breeze. Most people are intimidated by the thought of studying entomology—and particularly this branch of it—because the objects are so small. Only magnify them to the size of lions and tigers, and they become equally strong personalities. Oh, that doesn't go for all of them, but neither are all large mammals recognizable at first glance. The many species of African antelope, for example, always confused me because I had neither the occasion nor the time to concentrate on them. Field rodents likewise look alike until you have some reason to seek their intimate acquaintance. The various kinds of *Culex* gave me that sort of struggle at first. Blessed *C. univittatus* did have those leg stripes, but some of the others were even plainer. I came to recognize *Culex guiarti* by its greenish integument and relatively bushy antennae, while *C. zombaensis* was somewhat larger than the average *Culex*, had a darkish stripe down the center of its abdomen on the ventral surface, and—something that the books did not mention—was more sluggish in its actions than other mosquitoes.

The plainest *Culex* of all was so challenging that when the first one appeared I mounted it on a pin in order to study it not with my simple eyepiece but under the dissecting microscope. I ran it through Edwards' key again and again, always ending up at *Culex perfuscus*. The name itself, from the Latin for "indistinct," indicated the insect's unremarkable looks. Though I was sure, I was not sure, if you can understand what I mean. I knew that this was a case in which I must have Botha's confirmation, so I took the specimen to him on return to Joburg. He peered at it, then sat down at his 'scope, went through the key, and at last looked up approvingly. "You are right," he said. "This is a very difficult form, but you have come to the correct answer. I am glad that you did, because Jim Muspratt does not think much of Edwards' keys, while I have always felt they were excellent. Your identification of *Culex perfuscus* supports my opinion."

By this time I was talking to Qmba Ngwenya in gestures—at least I felt that I was getting through to him. But whenever a snag came up, I tried to get Jack Tschembene's help. However, he was exceedingly limited in what he could do. He spoke enough English for us to communicate on quite a few subjects. That was not the problem. It was a complete lack of educational background. I am sure that the mosquito catchers and Jack were all really very bright. Some of them may have been brilliant. But they simply *didn't know anything*. Pottie told me that the boys, who were familiar with the local witch doctor, believed we took those mosquitoes to Johannesburg to make "white man's medicine." In a long—excessively distant—interpretation of that view, they were of course right. I gained an insight into Jack's ignorance of the world one evening after dinner when we went out to look at a magnificent full moon that was just rising.

"The Russians have recently hit the moon with a rocket," I told him.

Jack considered this for some time. Then he commented, "That bad."

"Why is it bad?" I asked.

"They break it, and the pieces fall down on us."

Jack gave me some trouble with the dry-ice chest. It was not very good in the first place, by which I mean that the insulation could have been better. Dry ice evaporated from it more rapidly than it should have. Every time I opened the lid to put in a freshly prepared tube of mosquitoes, a certain amount of warm air was bound to enter, and that of course promoted a further but unavoidable loss of solidified CO_2. Many a time I could hardly sleep on the night before our little plane was due at the Ndumu airstrip with fresh blocks, knowing that the box now held only a few remaining chips at the bottom. Especially during summer, total exhaustion of dry ice might lead to a rapid rise of temperature in the box, and a week's work of collecting, identifying and preparing specimens might be lost.

But we should not have been operating on so narrow a margin. Dry-ice requirements had been carefully calculated at the lab, with a fairly generous, but not excessive, allowance for error. After all, our budget was anything but extravagant, and dry ice was expensive not only intrinsically but also as far as its transport on a chartered plane was concerned. What I had to fight constantly were non-entomological demands that both Pottie and Jack wanted to make on the box—*did* make whenever I was not on the alert.

I must admit that our meat had a rightful place in the chest, but that could almost be considered entomological because the entomologist and his assistants had to be kept properly nourished. We would bring a whole week's or two weeks' supply of frozen meat with us from Johannesburg. This would still be frozen by the time we reached Ndumu, so that putting it in the box with the dry-ice blocks didn't really account for much dissipation of CO_2. Each large package that Pottie took out for Jack's attention from time to time would be utilized without any of the thawed or cooked portions' being returned. For such temporary storage we used a small kerosene refrigerator (which was Pottie's pet and peeve at the same time, for it was always going out and he was constantly fixing it—almost weeping when it misbehaved and then patting himself on the back when he had corrected malfunctions).

But after naming meat and mosquitoes, I have come almost to the end of the dry-ice box's duties. The only other exceptions that could be made were in favor of occasional specimens that must be taken to Johannesburg for special studies, chiefly carcasses of birds, small mammals and snakes, or tubes of serum derived from them.

Pottie was the worst offender. He drank enormous amounts of water, and especially during hot weather he liked it to be cold. We had several plastic water bottles, each holding at least two quarts, that were supposed to be kept in the kerosene refrigerator. The trouble was that the mosquito catchers liked cold water, too, and when they would come to the back door and ask Jack for a cool drink, he could not refuse them, and soon Pottie would find that the kerosene refrigerator held only freshly filled bottles of tepid liquid. Thereupon he would furtively stow one of them in the dry-ice box to get it cold in a hurry.

Now anyone who knows anything about physical chemistry is familiar with the fact that water has one of the highest specific heats of all substances. This means that the amount of energy, in the form of heat, required to raise its temperature, is much more than would be needed to elevate to a similar number of degrees an equivalent amount of another substance such as sugar or sand. The process works exactly the same in the reverse direction, according to the law of conservation of energy. Pottie's water could not have been better chosen to put maximum strain on the dry-ice supply. For every degree its temperature was lowered, it gave up multitudes of calories that must naturally enable rebellious solidified carbon dioxide molecules to regain the atmospheric state in corresponding abundance. Much though I enjoyed cold water too, I put an immediate ban on the use of the box for that purpose. Pottie and I always got along together enormously well, but I think that on that one point we came close to a rift. In any case, I'm sure he cheated when I was not on hand.

As for Jack, he transgressed only once, and I forgave him because I felt he did not know it *was* anything he should not have done. I must preface this episode by talking about cattle—and a few cabbages and kings besides. As most people know, cattle are the wealth of many African tribes, since the animals can be used as a bartering medium.

In the Ndumu region, eleven cattle was the price of one human female in reasonably youthful condition. Naturally, under such circumstances, one did not idly slaughter a calf or a cow or a bull just to eat it. Each money bag was tended all day by small boys as it browsed the countryside, and the practice of conserving as many kine as possible—to compete in the wife market—had led to a great deal of quadrupedal overpopulation, overgrazing, and consequent erosion of the land. The cattle were, astonishingly to me, exceedingly fine in this semi-arid region of sparse grass and thorny brush. One wondered where they could find sufficient nourishment to keep themselves so well filled out and their mahogany hides so shiny and sleek. Bob Kokernot later informed me that there were two complementary explanations for the phenomenon—natural selection in the first place having weeded out individuals unfit for survival in this environment, and high nutritive value in the deceptively meager-looking vegetation.

All the time that cattle were being tended so solicitously, everyone was starving for meat. Mealies (which were grown here, too) are notoriously deficient in protein. Someone has written that the introduction of maize from the western hemisphere was one of the worst things that ever happened to the Africans, for when they found that it was so easy to raise that crop, they abandoned some of their former, more nutritive forms of cultivation, mistaking sheer bulk of the yield as excellence, when what they had previously raised in smaller quantity was better for them. Since they would not butcher cattle, they tried to satisfy cravings for flesh by snaring wild animals. On our game reserve this means of procuring meat was intolerable and had been prohibited. A resulting situation of tension now existed that I shall document in due course.

Elephants were uncommon at Ndumu, though they existed in abundance in Kruger National Park only a couple of hundred miles distant. Indeed, I heard of only one elephant during my sojourn at the field station. This was apparently a solitary "rogue," and since it had been doing much damage to crops, as well as terrifying the local human population, a professional game tracker was assigned by the Parks Department to hunt it down. This the man succeeded in doing, after many weeks of patient pursuit, for the deranged beast was wily as well as a threat. Since the huge crumbled carcass now represented no more than so much waste protein, permission was given to the natives to dispose of it as they wished. Their need for meat—as told me by one of the game rangers who was an eyewitness—was such that they went into a frenzy with their knives while cutting up the elephant, as likely as not cutting great gashes in each other's arms as all hacked away in the deep interior. It is a marvel to me, then, that they can hold themselves in such restraint when it comes to their own cattle. The event reminded me of people in our society who have been discovered dead of starvation on mattresses stuffed with fortunes.

(The young ranger, incidentally, removed a number of ticks from the elephant's ears and brought them to me. Dr. Zumpt later informed me that they were a specialized sort of tick found only on elephants. Their mouth parts were greatly elongated in order to reach capillary vessels beneath their host's thick epidermis.)

To get back to cattle and Jack, I must say further that bank robbery in Tongaland society was as prevalent as elsewhere, though the assets were not easily negotiable. Cattle branding was practiced, but brands can be altered, and animals could be driven for long distances to places where symbols might not be recognized. Cow-napping was therefore not unknown. But in addition there was another type of crime—that of revenge—which often led to a feast. If you had a grudge against your neighbor, you might go out at night and stab his best bull to death with an assegai. This was bank robbery just as fully as if the animal had been abducted—but at least you could eat the abandoned loot.

In more serene circumstances a bovine of whatever sex simply died of old age. That might happen out in the bush somewhere, and a convergence of vultures sometimes was the first clue to where the expired remains lay.

Murder or old age made no difference to the natives. Suddenly, here was available meat, as well as a hide that must be salvaged.

Jack had a kraal and two wives nearby. One day one of his animals was missing. When it did not return that evening, he set out on a search for it. Three days later he found it, assegaied some distance away. He enlisted the help of several friends to drag the body to our camp, and then he spent another day cutting it up and selling fragments to the entire populace, which had arrived for the occasion. By this time the stench was terrible (for it was summer), but no one seemed to think anything was amiss. Decay was apparently considered a precooking or marinating process. Some of the people ate their liquefying portions raw.

Jack covetously saved the best cuts for himself, and it was when I found that he had stored great flanks of carrion in the dry-ice chest that I really "flipped." That did not disconcert him overmuch. He looked more unhappy about having incurred my displeasure than upon being ordered to remove his treasured meat at once. We stored our viands there (he must have reasoned) so why couldn't he do the same? With easy grace he took out the offending chunks and hung them on nails on the wall of his unventilated shack, where they were quickly welcomed by clusters of flies and became ranker and ranker during the week that it took Jack to gnaw the last bit of gristle from the maggoty bones. I cannot imagine why those people did not die. Nor can I think what their taste buds could have been like. It must be all a matter of early training, for they seemed to think that it was all *good*. Or can it be that protein deficiency does this to you? I remember hearing that Welsh coal miners, who may lose six or seven pounds of sweat during one day's work, find water unpalatable unless it has been highly salted. All I can say is that African residents of Ndumu are misers when it comes to their own cattle, and in comparison Fagin or King Midas are only apprentices in the art of avarice.

I was fortunate to have come to Ndumu for the first time in October. That trip occupied dates from the sixth to the sixteenth. As I have said, early showers were already in progress; otherwise I would not have found *Aedes circumluteolus* so promptly. (The various species of *Aedes* are characterized by their laying drought-resistant eggs. In fact, some of the species *require* that their eggs be desiccated before hatching can occur. This is true of some other mosquito genera as well, though the more usual case is for eggs to hatch immediately after embryonic development. *Culex*, for example, and also *Anopheles* species deposit eggs on the surface of water which hatch within a few days, depending on ambient temperature. Should the pond or ditch run dry, those eggs would perish. *Aedes* females generally lay eggs not on water directly but in marginal situations such as moist mud or leaf mold, where water may be expected to reach them after future rain—though how they are able to select appropriate sites with such forethought remains unexplained. Consequently breeding cycles of *Aedes* species are characterized by long interruptions annually wherever pronounced dry seasons separate wet periods. At the end of four or five months of drought, flying adults can no longer be found in that sort of environment.)

Meanwhile development of *Aedes* embryos within their eggshells had already progressed to completion, actually within a few days after oviposition. It was the fully formed first-stage larvae that were able to aestivate almost indefinitely within their impervious shards. The ability of *Aedes, Psorophora, Eretmapodites* and similar groups to carry over unfavorable climatic gaps in this wise gave them a tremendous advantage over other kinds of mosquito. Poor old *Culex* and *Anopheles* had to go on breeding constantly or else become extinct. You could find their adults at Ndumu at all times of the year, though in the dry season they were mighty scarce. There simply were not enough suitable pools remaining to give them room for large-scale breeding. They resorted to a kind of resting state in shelters such as tree cavities and ground holes, but that was not nearly so effective as drought-resistant eggs would have been. In temperate regions some mosquito species have developed larvae that can survive winters in frozen aquatic environments, but their eggs remain tender.

Thus early rains bring out swarms of *Aedes* from last year's eggs at once, while *Culex* can take advantage of restored water reservoirs only by building up their populations painfully through several generations from the remnants of an erstwhile thriving community. *Aedes* eggs actually burst into life, and the first crop soon completes its aquatic cycle and takes to the air in full strength. The few surviving *Culex* females deposit their egg rafts sporadically here and there, an advent that is followed by only a negligible first generation of the new season. After several egg-to-egg cycles *Culex* begins to be conspicuous, but it may take them almost until the end of the rainy season before they can become pests on a scale to rival *Aedes*. The only disadvantage that *Aedes* suffers is that it is rain-dependent in the opposite sense. Until its eggs have had a chance to dry, they cannot hatch—indeed they may die if kept wet. Therefore intermittent rains are quite acceptable, but should precipitation remain constant, *Aedes* production may fall off. I could often tell the recent climatic history of Ndumu simply by looking at mosquitoes. Whenever catches of *Aedes circumluteolus* included a high proportion of bright yellow individuals, I knew that they must have emerged recently because a good downpour had drenched the area a week or so ago. When rains failed, the *A. circumluteolus* population soon became worn and drab-looking.

Don't imagine that I was not looking at birds all the while. During this first trip to Ndumu Bruce and I went out every morning to admire them. I had at least two free hours after breakfast before the first box of mosquito tubes would arrive, and Bruce felt he should aid me in other phases of natural history. As a matter of record, he pointed out thirty-five new species during this ten-day session alone, though eventually I picked up several times that number at Ndumu. Of these there is not much I can say unless I were to tell only of birds, whereas I am now wearing the robes of an entomologist. However, apart from Emerald-spotted Wood Doves and their melancholy cooing, the Purple-crested Turaco, Trumpeter Hornbill, White-eared Barbet, Scarlet-chested Sunbird, Firefinch, Martial Eagle, Orange-breasted Bush Shrike, and the iridescent green male Klaas's Cuckoo fairly shriek to be mentioned as trophies on my list. Their names alone are sufficient to signify how good Bruce and nature had been to me.

Breaking camp was more simple than I would have predicted. The food had all been eaten, for example, so we did not have to worry about that. Or, rather, such items as remained were not worth carrying back and were eagerly received by Jack and the mosquito boys. The latter, of course, had had to be paid, and Pottie made a great ceremony of that transaction on the afternoon before we left, counting out so many shillings and pence to each one and entering the amounts in a ledger. The smaller boys received less than older ones, but no one raised an argument after Pottie gave out what he considered to be the proper dole. Undoubtedly these Africans' ability at arithmetic was not of a high order, so that it would have been difficult for them to quibble or bargain, had Pottie short-changed or consciously underpaid them. Indeed, he himself would come to me sometimes with his ledger and ask me to "tot up" his figures. "I've done it three times, Doc," he would say, "and it never comes out the same."

Whenever we put on a "session" at Ndumu, Dupe knew what he was in for. In the first place he must get in a large supply of change, for the boys all wanted to be paid in coin, and Pottie went to the store a few days before our departure to get a whole bag of coins in lieu of pound and ten-shilling notes. Then, after the shillings and pence had been counted out, the boys would all clamor for Pottie to drive them to Dupe's so that they could spend it. Thus the money all came back in short order. The youngest mosquito catchers—Magwalo, for example, but also Qmba Ngwenya—hoarded their few coins to give to their mothers. Somewhat older ones spent part of their pay on trinkets and novelties such as mouth organs, tin whistles, and Superman sweatshirts. The most sophisticated ones would buy gaudy cloths of various sorts that they said were for *their* mothers, but one could well imagine they were for nubile mothers-to-be. All of them were mad for soap and for a certain brand of furniture polish which, after about three days, turned their hair an odd reddish-straw color. Everybody was gay, and they all treated Pottie as if he were Mr. Rockefeller himself. How could they know otherwise, as he handed out largess to one and all? Catching mosquitoes was really nothing. It was a "ball," as we would say these days. For having had fun while doing inconsequential work, simultaneously being relieved of the monotony of tending cattle, they had been paid hard cash. Hail to the white man!

In late afternoon we loaded the van except for last-minute things such as blankets, suitcases, and thermos flasks containing dry ice and the most recently collected mosquitoes. Then it was time to sit around with Medical Reserve. Shortly, Jack called us in for leftovers, which nevertheless included great beefsteaks, so that we went to bed early on distended stomachs. Three o'clock in the morning seemed not so grim as usual when Jack brought heavenly hot tea to our cots. The drive, however, was again harrowing. Without having to interrupt our journey as at Germiston on the way down, we made much better time and reached the lab by one o'clock.

Ken practically embraced me. "You *did* it," he exulted. "I knew you would. Well, I hoped you would. We've got one mouse group that already looks suspicious, from an early lot of *circumluteolus* you sent. But even if that does not pan out, we've got mosquitoes coming in again, and that's what we needed. I couldn't be more pleased."

Great praise from a great man! *I* couldn't be more pleased either. I spent the afternoon in my office, studying a few pinned mosquitoes that had confused me at Ndumu, but I was so weary that I did not concentrate well and kept looking at the clock to see when it would be permissible for me to take off for the Garden Inn.

The hour arrived. I shuffled to the parking area with my suitcase and unlocked the Renault. Rosebank was four or five miles distant, one stretch of the road being an unpaved, dusty shortcut that always gave my little car a bad cough. As I was making my way up the farther slope, Ken reached me in his Chevvy as if this were a superhighway. He braked his vehicle while passing to call, "Come up to my place and have a drink." The very thing! Why yield to fatigue when there are ways to combat it?

"OK," I shouted, "I'll be there soon."

6

KEN, BOB, PAUL AND HUGH

Ken lived as a bachelor, his wife having returned to the States from Entebbe, Uganda, for reasons of health. His consequent loneliness augmented his natural hospitality, though I flatter myself by thinking that he would have warmed to me anyhow (that is, as long as I had proved myself able to identify African mosquitoes). He now asked me to sit in the living room of his apartment while he mixed a pitcher of Manhattans. I am not much of a one to sit when I can prowl around someone's private haunts for the first time. What books, magazines, pictures, knick-nacks have been hoarded by the unfamiliar individual?

I was at once struck by the lack of resemblance of appurtenances in this room to those you would expect to find in a virologist's lair. Though we did ultimately talk about viruses (an inevitability when two such as we were not beholden to other company), I could see that Ken had consigned his scientific life to the lab but had reserved a large measure of his leisure to other types of enjoyment. Most of the current novels were on his shelves. The place abounded in art objects of African origin. On the walls were framed paintings, most of them prints, but a few original oils among them. Most of the prints I could recognize. However, there was one that puzzled me because it did not seem to be complete.

It was a vignette, excerpted from a larger canvas, depicting a drunken scene at a rude wooden table, with spilled mugs everywhere and one of the celebrants out cold on the bare ground.

Ken entered with a tray holding glasses, the pitcher, and a plate of canned smoked oysters. Soon we "cheered" each other, and then conversation began to well from Manhattans' springs.

"I have been admiring your paintings and prints," I said, "but there is one I want to ask you about. This one over here is very funny. It looks to me like a detail from one of those vast outlays of Pieter Breughel's that depict panoramas of children's games and so on. I would imagine that this is a similar scene involving peasants' pleasures. They are almost like cartoons, and at the same time something like modern primitives. But what am I saying? Can you tell me what the picture is?"

Instead of answering, Ken marched to the wall, removed the frame and showed me the typed legend on its back: "Detail from painting by Breughel the Elder." He then licked his thumb and executed a pantomime of wetting a scoreboard with it. "One up for you," he said. "Now tell me who did my originals."

I confessed at once that although I had already studied them I could not ascribe them to any source. "They are by my wife, Florence, now living in New York," said Ken. "She did them when we were in Entebbe. What else you see on the walls here are our favorite choices among the classics. She has always been an enthusiastic artist."

So science and culture were merged in this man on a family basis as well as by his own inclination. For the hundredth time I decried the view that science can be held to be something *other* than culture. A person who is really alert to the verities of science must be equally susceptible in other spheres of appreciation.

"When we were in Entebbe," said Ken, while pouring our third drinks, "Sandy Haddow and I worked on viruses together. He and his wife would come over to see us on one evening, and we would visit them the next. There weren't many other people around. Sandy was a Scotsman and very brilliant as a general naturalist as well as virologist. Of course I don't know much about birds, but at least I know a robin when I see one.

"One night, after we had had quite a few drinks, Sandy happened to begin talking about robins, in passing referring to them as tiny birds the size of wrens. 'You can't be talking about robins,' I said. 'Robins are big birds. I know, because I grew up with them in Indiana.'

"We got into a great fight, and I finally ordered him out of the house. (They were visiting us that time.) Naturally, we had been arguing about two kinds of robin, and I'm sure we both realized it soon after the hostilities broke out. But by that time we were both too stubborn to admit it. Next day each of us pretended that he had forgotten the dispute and my shocking behavior—though Sandy had been quite nasty, too."

At any rate, I had now "made it" with Ken, not only in the entomological arena but also in the realm of painting, though that Breughel had been only a happen-so: Ken's knowledge of art was far more extensive than mine, and had he had some other picture at that spot on the wall, I could not have made what seemed to him an informed comment about it. But you may as well accept luck when it comes your way. Any number of times thereafter he passed me on the dusty piece of road on the way home after work and hailed me with further invitations to join him in "cheers" and smoked oysters.

Then Bob Kokernot arrived back from leave. Ken's shortness of breath had been plaguing him increasingly, and he welcomed this inexhaustibly energetic colleague with more than ordinary thanksgiving. Bob virtually took over most of Ken's duties, but not without another set of rules: oh, no—Ken reigned over procedures in serology and virology as rigidly as he regulated my actions at Ndumu.

Bob soon occupied as large a portion of our lives as his exaggerated map of Texas, sent to James H. S. Gear, had presaged. I was immediately attracted by his fresh handsomeness, which evoked long days out-of-doors in the saddle, as well as by the warmth with which he accepted me as a new colleague. He held only two doctoral degrees at that time, in veterinary and human medicine. His third doctorate in public health was yet to come. Texas is not only three times as big as the rest of the world: you have to have three times the ordinary quota of degrees to qualify as a citizen of that focus of our Union.

But let me hasten to add that Bob Kokernot was not only three times as ambitious as we lesser beings from the Northeast but three times as handsome and nice, to boot. In other words, everything about him was in proportion, albeit the slices were large. He absolutely exhausted me on occasions when he felt he must come to Ndumu for some reason or other. He worked like a dog, but afterwards played like a puppy. Prior to his advent, he would tell Pottie to assemble as many natives as possible on a designated day so that he could hold a "fever clinic." That was no more than a blood-letting session, during which he emptied veins of their contents into Vacutainer ampoules in order to keep a serological check on the most recent activities of viruses in the region. "Patients" who had been bled previously could now be tested for "conversion" to various viruses. If stored serum samples from each named individual were negative but currently taken samples were now positive to a particular virus, such conversion proved that the specific disease agent had been active in the interim—an item of information of prime epidemiological importance.

The trusting natives thought they were being treated for whatever might be wrong with them. Of course they visited their witch doctor, but he did not possess such beautifully shiny instruments, and in any case he was rather austere. Bob absolutely charmed these people and they adored him. Furthermore, they believed that the loss of blood was ridding them of poisons, and the sugar tablets that were administered afterwards must certainly contain further cures beyond the scope of the medicine man's incantatory art.

Little children were remarkably stoical about being bled. Indeed, that was a disadvantage in some cases. Infants under the ages of three or four have extremely small veins in their arms, and it is often impossible to find them with an exploratory needle. Therefore the external jugular vein must be probed. This vessel becomes moderately distended, so that it can be seen faintly, if the infant is held in an inverted position, and Bob would instruct the mothers to hold them in that stance. This maneuver, plus the threatening presence of a blustering white doctor, should have thrown any self-respecting infant into a hysterical panic, when weeping and screams would have bugged out the external jugular like a garden hose. But not these stolid babies. They hung upside-down between their mothers' knees as if this were an everyday nursing position. Bob sometimes pinched their earlobes (on the opposite side from their mothers' vision) trying desperately to make them cry, but even that worked only occasionally.

The clinic dismissed, Bob would be ready for a Texas spree of some sort. What should it be this time? Perhaps at the last Ndumu session it had been a barbecue, so naturally that could not be repeated so soon.

"Yahoo!" he might shout. "Let's all go to Catuane!"

Everyone would jump into the Land Rover and drive to our most distant collecting station on the banks of the Usutu (or Maputa, as the Portuguese call the river), about ten miles from our field headquarters. Here a ferryman would row or pole us across illegally (since we did not carry our passports in the sticks), and then we walked a further mile to the tiny but magical village of Catuane—magical because here was the closest bar to which we had access while sojourning in the wilds of Ndumu. However, that wretched Portuguese Constantino brandy made it scarcely worth the effort, in my opinion, to go through all the clandestine procedure to get there. The beer was good, I'll admit, but chiefly because it was cold, this little store being favored with electric power wired down from the capital, Lourenço Marques, for running refrigerators.

It was, of course, the outing itself that made the excursion worthwhile. However, on one occasion when Bob shouted "Yahoo!" and David and Bruce rose to follow his lead, I declined. I had identified over a thousand mosquitoes that day and was more than tired. Besides, I quickly realized that if Bob and his fellow revelers went to Catuane, I could get to sleep early instead of lying next to bedlam until some late hour. They all tried to change my mind, but I remained firm.

That is not the way to play with a boy from Texas, as it transpired. They left, and I went to bed, and everything seemed rosy. God knows what time it was, but suddenly an explosion took place under my cot. That damned store sold not only beer and Constantino, but also enormous firecrackers. I roused my trembling frame and had a now needed nightcap of Medical Reserve with my friends as they finished laughing at their devilishness.

Barbecues were not quite so impromptu. At least the first one had not been. Some later ones came to Bob as inspirations from who knows where, and he might be off at ten o'clock at night in search of a young calf or goat he had all at once taken into his head to require. He never came back without one, but whether he had stolen it or murdered its owner he would not say. However, the most elaborate occasion was preceded by a round of conventional invitations to everyone in the Ndumu region: Dupe and his rarely seen ill wife; Oppie, the cattle-dipping inspector from Ingwavuma; personnel of the tobacco farm; game rangers; and other officials whom I forget.

In all there must have been thirty people that particular time. Bob had the mosquito boys dig a pit alongside the house, and then he found an iron lattice somewhere to use as a grill. He spent all day basting the calf with a floor mop that he kept dipping into a bucket of ketchup, Worcestershire sauce and other condiments. Eventually the carcass became utterly black in its casing of charred ketchup, whereupon Bob said it was deliciously ready. I suppose my trouble is that I don't come from Texas, for I thought it was ready to be thrown into the delicious Usutu River.

Paul Weinbren, a Johannesburger, had been with the virus unit at its beginnings, but had then gone to work for a couple of years

with Alexander Haddow in Uganda. There he had made a stimulating study of the possibility that a species of striped field mouse
(or rat), *Arvicanthis abyssinicus*, might be a reservoir for Rift
Valley fever virus, heretofore known to affect only sheep and
human beings. But Paul was—and is—in no sense a mammalogist,
versatile though one and all must acknowledge him to be. Apart
from his preeminence in radiology, photography, electronics and
gadgeteering in general, he was primarily a virologist, as far as
ABVRU's requirements of his talents were concerned. Thus, when
Paul rejoined us, that put him, Ken and Bob all in the same slot,
and ABVRU might look a bit top-heavy in the virology department. That was not really the case. Virology is so intricate a subject that you need lots more brains at that level than are required
when it comes merely to chasing insects.

Of course I must retract the last clause immediately. After
Hugh Paterson's return from England, I was suddenly possessed
of a most erudite entomological colleague. My office (which had
been *his* office) now rang with wonderful conversations on every
biological topic you can imagine. Hugh seemed willing enough to
fill me in with recent academic lore from high places. He was a
tall, deliberate young man who prefaced answers to anything I
said with the drawn-out words, "Oh . . . yes." Then he might go
on to show me where I was wrong.

I had wondered what would happen when Hugh came back. It
seemed to me that ABVRU did not need two entomologists. By
this time I was in love with Ndumu and its mosquitoes, and competition might have led to difficulties. I needn't have worried.
Hugh had a young wife and did not want to go to Ndumu on
extended trips. He could identify mosquitoes, of course, but he
was much more interested in rearing them in a laboratory insectary than in labeling their dead remains in a hinterland. Indeed,
he preferred houseflies above all other two-winged insects. Finally, he still had his doctoral thesis to prepare, and if I wanted
Ndumu for myself, he was more than willing to concede me the
honor. We did not clash in a single sector, and his "Oh . . . yes"

was always so delicately modulated that I rarely appreciated its negative import (until in bed at night).

Ken, managing a team that included a Texan, a Pennsylvanian, and diverse South Africans, tore his hair at times. That was natural. Any director has to do that to his toupée, but I am sure that this was not a factor in the progressive deterioration of his health. His shortness of breath (first mentioned to me in Grand Central Station) became worse until everyone was alarmed, last of all Ken himself. He kept asking me to his flat for a drink after work, but he no longer passed me at the same spot on the dusty shortcut, and I would sometimes get there first. To walk to his car in the parking lot at the lab became too great a chore; he now delegated Pottie to bring his Chevvy to the door.

We had a magnificent pre-Christmas party at the lab that first year. It was hardly an "office party," for we held it in the carpenters' shop. All tools were cleared way, and benches and stools were brushed clean to make room for glasses, ice buckets, bottles, and places to sit. Ken was the master of the occasion, naturally, and we all became distressed when, after the very first drink, he was commandeered in conversation by Mr. Kruger, our head storekeeper—a most prosaic individual who became red-faced and infinitely voluble about nothing at all the minute he touched a drop. I was finally assigned to take his attention away from Ken by fabricating interest in the nonsense Mr. Kruger was talking, but that did not work. Ken felt that the man should be taken home, and only he would do it in order not to spoil the fun for anyone else. Off they went. A long time passed, and we were about to set off in various directions to find Ken when he returned. He said he had been delayed because Mr. Kruger had wanted to stop first at a supermarket to do some shopping and next to drop into a bottle store for a bit of whiskey. Dear Ken. Any of the rest of us would have told Mr. Kruger what he could do.

Not long thereafter (though not in consequence of the Christmas party) Ken entered a hospital in Johannesburg and was told that he had emphysema. At this mile-high elevation he could not get enough oxygen into his system to enable him to function, so there was nothing for him to do but return to his boyhood environs in Indianapolis. We had a final look at the prints and paintings in his apartment, knocking back Manhattans and smoked oysters, and away he went. Now he and Florence are watching birds (including robins) from their attractive Indiana home.

Ken's wit must be documented at least once, though I could go further. When he was in Uganda, government regulations prohibited any employer from issuing an unfavorable commentary anent a worker who had been dismissed for incompetence, dishonesty, or any other reason. Indeed the employer was bound to give a *positive* recommendation, no matter how poor the individual's performance had been. Unfortunately the literacy rate was high in Entebbe, so that the natives could read "chits" they received at the time of their discharge. Therefore Ken could not count on his sealed letters' not being read by their recipients. Some idioms, however, are not covered by grade-school readers. He gave a note to an especially inefficient animal attendant (who must have been elated by its contents) that read, "If this man asks you for a berth, give him a *wide* one."

7

POLLUTED WATERS

Johannesburg itself sat atop a vast catacomb of mine shafts. Many a night at the Garden Inn I would feel a rumbling and think that my rondavel was about to collapse into a cavern below. In the morning I was told that it had been only a settling of timbers in an abandoned tunnel a mile or so underground.

Modern gold mining required the constant evacuation of subsurface water. With effluent from gold refineries added, this flux fouled the only natural streams around Johannesburg. Recognizing that fact, the authorities did their best to impound polluted waters in artificial lakes, or to direct them where they would be least harmful to wildlife. Yet the volume was so great that I did not find a single brook or rivulet that lacked a characteristic sweetish, chemical and fetid smell, all in combination.

In a quite remarkable way, acid residues and other waste products fostered a teeming horde of invertebrate life, for example, insect larvae and small species of shrimps, that in turn invited some vertebrates such as birds to feed upon its bounty. Where concentrations of pollutants were strong, only a few higher forms found the environment tolerable but, strangely enough, the Greater Flamingo was one of them. This bird was able not only to withstand corrosive effects of mine waters on its shanks and feet as it waded in the evaporation basins, but could also immerse its beak with impunity in order to seine out spineless water creatures. The Lesser Flamingo, not much smaller than the Greater and strikingly resembling it, nevertheless differed widely from its congener in a physiological sense, since it was a vegetarian, subsisting chiefly on algae. Whether it found mine waters equally salubrious I do not know, but my guess is that tiny invertebrates could not have flourished in that environment unless algal "grass" had been there to feed them, and therefore the Lesser Flamingo ought also to have been an ultimate beneficiary of mineral concentrations that were the basis for this abnormal but apparently rich and successful food web. The smaller bird was less common even in more natural habitats, and that is perhaps why I missed seeing it in those pungent ponds.

One of Bruce's first acts, after my arrival in Johannesburg, had been to introduce me to the Melrose Sanctuary, within walking distance of the Rosebank community. Here a rather small stream, with only a moderate degree of chemical contamination, had been dammed to form a miniature lake with a surface area of not more than an acre or two. Yet in the hilly precincts of Johannesburg horizontal expanses of water were so scarce that even this tiny haven was frequented by more aquatic fowl than it had means to support. At least that is the impression I received, for there was constant bickering and chasing among coots, grebes and ducks. Red-knobbed Coots were the most aggressive, toward their own kind as well as other species. They nested among the reeds, even directly in front of an observation point where the public assembled to view the entire panorama. Whether spectators were present or not, when it was time for copulation a female coot would clamber shamelessly to her partially finished nest so that the male could do likewise and mount her. Either coots cannot cohabit in water or their resort to nesting sites is a territorial act that maintains bonds between mated pairs. We have trouble telling one coot from another, and maybe they suffer from the same difficulty. A nest could therefore substitute for a latchkey or for a number on a house that resembled every other house up and down the street. Whatever the case, female coots were advocates of sex strictly at home.

Everything I saw at Melrose was new to me except Little Grebes and Common Mynas. Those were Indian birds that I knew well. Mynas had been brought to South Africa and liberated at Durban in Natal province—or perhaps they had escaped—but now they were establishing themselves as firmly as our English Sparrows and Starlings at home. Otherwise the coots were novel, despite still being coots, in possessing two red excrescences atop their white frontal shields. As for the ducks, these, too, had counterparts in my experience. You got a clue to the Cape Shoveler when you saw its expanded, flattened bill, and the stiffly spined tail of Maccoa Ducks disclosed their affinity to our "Ruddies." But plumages tried to belie relationships, and I felt I was the butt of several kinds of ornithological joke.

In more exotic departments there was no question of playing superficial games. Sacred Ibises, though accorded that name because of the veneration in which ancient Egyptians held them, are by no means confined to the Nile Valley and often appeared at Melrose, either stalking snails at the edge of reed patches or gleaning invertebrates in nearby fields. You don't get a proper idea of their elegant appearance from Egyptian carvings, which are highly stylized. In particular the long black plumes of the lower back can not be rendered adequately in stone, though the rest of the silhouette is simple enough. The idea that these ibises were sacred has been interpreted by modern epidemiologists on the basis of their snail-eating. Aquatic snails are intermediate hosts for blood flukes that infest man and other vertebrates, causing a serious disease variously called schistosomiasis or bilharziasis. Dissections of Egyptian mummies have disclosed eggs of those flukes in the abdominal viscera of Pharaohs dead for thousands of years, though clinical evidence can no longer be adduced to say whether that is what killed them. Many people in the tropics today get along quite comfortably with light infections. But schistosomiasis *can* be fatal, today as in former times, and it is supposed that the Egyptians observed that where ibises abounded the disease was absent, or perhaps only mild. The more ibises, the fewer snails (though they did not realize that facet of the epidemiology), and the more safely could rice cultivators tend their inundated crops.

At Melrose the flocks of those birds, largely white with long curved beaks and naked black heads and necks, were always a treat, for you could not count on their presence (as you could on the grebes, coots and most of the ducks). Where the ibises went at other times I have no idea, but when they were on hand they behaved as if this were a chosen home. I concluded that they did not wander far, but I could have been immensely wrong on that score. I hope that the Johannesburg Bird Club will undertake to "ring" this flock (which must still exist) to settle the question. If one of the marked birds is ever found dead at the foot of a pyramid, I offer to eat it—the bird, that is.

When one sat very still almost anywhere along the banks of the pond, one often saw small dark figures moving furtively on the ground among stalks of vegetation. They were hard to see, and at first I imagined they must be some kind of rail, for the habitat was perfect for such birds. However, I was eventually able to get a good look at one through binoculars and was delighted to find that it was a dark-colored, short-tailed mouse—unless it should be called a rat. Here was another example of a diurnal rodent. I don't know what advantage they found in that reversal of the usual nocturnal habits of rats, for Blackheaded Herons, being diurnal also, must have been a far greater threat to the species than enemies at night could have posed. The so-called vlei mouse—a species of *Otomys*—that I now watched at last came fully into the open to feed on succulent grass stems. Another departure from conventional rodent behavior. What were its cutting incisors for, if not to gnaw on seeds, nuts and grains? A heron stalked toward the mouse, and I got up and shouted. One took to the air, the other darting into the blackness of the reed bases.

A real local oddity, not too unlike an ibis, was the Hamerkop. Translated from Afrikaans this meant "hammerhead," which is what English-speaking taxonomists call it. The bird has no close relatives and is placed in a family by itself. The plumage is an unremarkable dark brown. But, as the name suggests, its head gave a most unusual impression of being beaked both fore and aft. The moderately long straight bill had its counterpart in a crest of almost exactly the identical shape, lying backward rather than erect, at the same angle from the skull as the bill. If it had turned its head to examine the rear view, you could not readily have told the difference.

The stream leading to Melrose Sanctuary probably contained other contaminants in addition to mine water, for the surface of the pond itself was sometimes flecked with masses of spume that collected there. I took these accumulations to be detergent suds. Perhaps there was a commercial laundry somewhere above. The foam converged in masses at the dam at times, or, if the flow were strong following a rain, it cascaded over the brim and then formed even greater mountains of froth after its churning ten-foot drop.

One could walk part way along the dam until one reached the spillway. Here, on the lower side, sprang a great willow tree, and in a crotch not much higher than my head as I stood on the elevated masonry was a Hamerkop's nest. The classification of birds is based mainly on their physical structure, but occasionally taxonomists recognize exceptional behavioral traits as further guides for interpreting genealogies. The Hamerkop's nest is an example that prompts the latter practice, for it bears no resemblance to that of any other avian group with which it might otherwise be considered to have affinities. Storks, ibises, herons and other such birds all build shallow, flat-topped nests that well fit the concept of a "generalized" nest, if there be such a standard for nest construction. Certainly that is a very simple and primitive form—the first step beyond making no nest at all and laying eggs simply on the ground or in hollow trees. On the other hand, mud nests of swallows and woven penduline nests of many other passerine birds fall into the sophisticated architectural category.

The object I was observing in the willow tree looked like a dense pile of trash that had somehow become lodged there as if in a great flood, although the little pond had never reached such heights. This object was made of sticks, leaves, grasses, papers, scraps of cloth, and anything else the birds had been able to find and transport. In shape it was almost spherical if you discount the ragged tatters that hung from its sides and bottom. I judged that it was about six feet in diameter and almost as deep, which, according to the formula for the volume of a sphere, would mean that the birds had collected at least one hundred cubic feet of material to build this home. Roberts reports that it takes Hamer-kops six months to fabricate each nest, and that Barn Owls and honey bees sometimes preempt the finished structures so that the birds have to begin all over again. The nesting cavity in the interior is reached through an opening from below.

On the other side of the dam wall from the Hamerkop's nest, a bed of reeds was home to a colony of Red Bishop Birds, members of the Weaver family to which our English Sparrow belongs. The females looked sparrowy enough to be acceptable members of the tribe as we are accustomed to think of it, but the brilliant black-and-red males came as a shock to me at first. (Later I saw many other equally spectacular kinds of Weaver and became used to the idea.) Red Bishops are polygynous, each male accumulating a retinue of two or three wives for whom he builds just enough of a nest to lure them into his harem, whereupon he allows them to finish the job while he goes on his next amorous quest.

Kingfishers were there, too. In fact almost every type of water bird showed up eventually, so that mine wastes and detergents must have had a singularly mild effect on aquatic ecology. This is still as incredible to me now as it was then. Giant Kingfishers, a foot and a half long, always delighted me simply because they *were* so big. Had I grown up with them, I would have found them ordinary, I suppose. When, at a much later date, I reported to

Bruce that I had seen a Half-collared Kingfisher (of conventional size) along the stream below the dam, he was the one to be emotionally aroused, for that species had as yet eluded him. Both of us conscientiously kept our life-lists up to date, so that we each knew what birds we had already seen, as well as those species that remained to be encountered for the first time.

Melrose was always good for an hour or two when one did not have time for a more distant venture. But Bruce did not terminate my indoctrination there. Scarcely ten miles or so from Joburg were two collections of impounded water of much greater size, though still nothing to show on anything but a local map. Ronder Vlei duplicated Melrose's smell of mine water and show of detergent foam, while Olifant's Vlei had the distinction of receiving effluent from a sewage-treatment plant. Again, birds and sub-aquatic creatures abounded at both places. I am not trying to establish an argument in favor of pollution. However, it seems to me that we can see evidence in these examples that the *types* of pollutants must be carefully specified when criticisms are raised against them. Chemicals that are toxic rather than nutritious would have produced desolation instead of these thriving natural communities.

I made several professional trips to Olifant's Vlei with Adrian Boshier, one of ABVRU's entomological assistants. (Olifant is obviously the Afrikaans name for elephant, while a vlei, pronounced "flay," is a body of water with flanking marshes.) Adrian had keen eyesight and could catch mosquitoes when asked to. We were interested at ABVRU in a Highveld mosquito, *Culex rubinotus*, that had recently been found carrying an obscure virus, and my colleagues wanted to test more of them. Despite our searches, Adrian and I could never find additional specimens. Perhaps that was because my alleged assistant was more on the lookout for snakes than for insects. I didn't see any snakes, either, but he was absolutely phenomenal in that department. I could hardly bring the car to a stop at Olifant's Vlei when he would be out of it and have the first serpent securely tied up in a cloth sack.

Once one traveled to greater distances from Johannesburg, especially in the upstream direction, water collections became sweet, in the sense that they had no odor or taste. For example, Bruce showed me Hartebeestpoort Dam on the way to the Magaliesberg Range, a fine large reservoir about forty miles away, that was used also for sailing and swimming, though it supported snails capable of hosting bilharzia flukes and was not a place I would care to use for bathing. Fish Eagles, resembling our national bird except for rufous shoulders and underparts, could often be seen there. Somewhat closer to town he made a stop along Hennop's River, where Johannesburgers often came to camp in mild weather, and I was rewarded at once by the sight of a Red-billed Woodhoopoe—a strange bird with woodpeckerlike habits but a most unwoodpeckerlike appearance, being slender of bill and having dark iridescent-green plumage.

One of the best trips on which Bruce took me—and almost all the ones I have mentioned were during my first six weeks at the lab when I was studying pinned mosquitoes—was to Barberspan, west of Joburg by a good hundred miles, if not farther. Most of our expeditions were taken on weekends and for fun only, because Bruce clearly enjoyed pointing out new birds to a tyro. I know precisely how he felt, for I have done the same thing elsewhere for foreign visitors with ornithological leanings. When English Sparrows and robins have become commonplace to you, and particularly when you are so jaded that even fairly rare birds are no longer novelties, the only way you can recapture a sense of your original emotions on having identified them is to witness the rapture of discovery as it flushes the countenance of a less jaded friend.

Barberspan represented fun, to be sure, but the basis of that sally was business. This was an exceptionally large pan (or lake), though quite shallow over most of its extent. Consequently it attracted hordes of wading water birds in addition to some diving species. I believe that the area was a sanctuary, or in some way under the jurisdiction of the government game department, for otherwise it would long since have been blasted clear of the last duck.

At that particular time a young European ornithologist had come to Barberspan with his wife and new infant to make studies of the birds, either under government auspices or by virtue of a grant from his own country. He was a Czechoslovakian, or possibly a Hungarian, with virtually no knowledge of English, so we never could find out what he was really up to, or who had contrived to bring him there. At least he loved birds and was trapping and ringing them at a great rate, while his isolated and neglected little wife could do no more than wring her hands in unhappiness. His chief interest was in ducks. He had rigged up a battery of large wire cages, with funnel entrances, at the water's edge, and these he baited with mealies, or whole-grain corn. Overnight many birds, including coots as well as ducks, would enter to feed, and in the morning the ornithologist and his local African assistant would remove them in sacks, carrying them to a small shed nearby where they could be weighed, measured, examined for stages of molt, ringed and released.

When Bruce heard of this operation, he thought it would present a splendid opportunity for him to collect a large series of blood specimens from an avian population living close to mosquitoes. Under ordinary circumstances you might spend a year hoping to bleed a wild duck and not have the wish come true, simply because you did not have time to become a duck catcher. Some arthropod-borne viruses were known to be "bird-associated," as the expression goes, but in most familiar cases the birds were not aquatic. However, water birds had not been canvassed nearly as thoroughly, as I have just indicated, and the Barberspan setup now looked like an ideal situation to exploit on an antibody survey basis.

Please don't ask me what the scientific results may have been. I suspect they were largely negative, for I ought to retain a less blank impression if that were not the case. What I remember vividly is the miserable hotel where we spent the night and our early morning arrival at the shed where sackfuls of birds already lay on the floor.

Bruce unpacked his bleeding equipment, which included needles, syringes, vials, alcohol sponges and a variety of paraphernalia, and the Czech (if such he was) rather unhappily passed ducks and coots across the table after he had finished his own documentation of their biological status. The African assistant, who could speak a mite of English, received the birds from Bruce after they had been bled and released them by the simple expedient of throwing them out the window.

In most cases there was a loud beating of wings and the weighed, measured, ringed, bled bird soon found a safe place to alight in the middle of the pan. But sometimes there was nothing but a thud. The Czech then looked disconsolate, as if Bruce's mutilation with a needle were fully to blame. Undoubtedly sometimes it *was,* but I have trapped and banded enough birds to know that those procedures alone, without bleeding, occasionally lead to unhappy ends. Whether the creatures have been physically injured internally in the processes of being trapped and handled or whether they respond to fright with reactions simulating apoplexy or heart attacks I can't say, but they do either "come over queer" temporarily or die outright without external signs or trauma or disease.

The thuds were not always death knells. Some of the ducks seemed only to have gone limp, as if resigned to being mauled. Thus, when the native projected them into the air, it did not occur to them to flap their wings. They might then lie on the ground for a few moments, as if the breath had been knocked out of them, then come to their feet, take a few steps, and rise at last into normally strong flight.

The ornithologist was greatly concerned about those birds that never raised their heads again. Bruce's needle, or some other cause, had done them in, and they were irretrievably dead. A few others began to make it back to the pan but collapsed on the way, either waddling or in feeble flight. The ringing records must be corrected in the case of each death, in order that the size of the surviving population be accurately known.

The native, with his limited English, gave us a countdown after each release from the window. "Flying all right," he said in most instances. Otherwise it was "Not flying." Bruce and I adopted his terse pronouncements in a way that he would never have imagined. Sometimes we added question marks. On mornings at the lab, I might say, "Flying all right?" and he would answer in the affirmative or else with "Not flying." At cocktail parties we were always flying all right by the end of the evening.

8

THE NDUMU GAME RESERVE

Game was scarce in the Highveld. At Barberspan Bruce and I saw a springbok, a type of antelope, though it was half tame and I did not feel that it was a fair example of a truly wild one. When I first learned that our virus field station was situated in the Ndumu Game Reserve, I formed visions immediately of vast herds of large mammals. What really happened was that I gained familiarity with several species of engaging rodents but hardly ever saw anything larger than a mongoose.

I should first excuse this lack of teeming herds by explaining that the Ndumu Game Reserve had only recently been established and was, in fact, still more of a concept than a reality (except on government papers). As I understood the situation, parts of Ndumu had been designated a game reserve only because a herd of hippopotamuses lived at the confluence of the Usutu and Ingwavuma rivers. It was not an especially large herd—no more than two hundred animals, I believe—but native hunters were giving them a hard time, and the authorities decided to save the beasts from extinction because no other hippos occupied this part of South Africa.

Hunting methods had been spectacularly inhumane. The African native is probably not wantonly cruel, but if he needs to capture meat, he must do it by whatever means he is able to contrive. He does not possess high-powered firearms, and his slender assegais would scarcely be effective against an animal weighing a ton or more. Of course he knows the habits of all kinds of wildlife. Therefore he takes advantage of what each kind of beast ordinarily does.

Two methods were employed, each one based on the fact that hippos spend their days in the water but emerge onto dry land at night to feed. Well-marked hippo trails become established at such points of emergence, and here the snares were laid. In the first case, natives would prepare a log, perhaps fifteen feet long, with embedded, sharpened spikes protruding from it. I presume that the spikes were barbed as well. The log would be placed at the center of a hippo trail after the animals had come out to feed, and natives would then hide in waiting nearby. At daybreak, as the procession returned, some hippo or other was almost bound to step on one of the spikes. These animals' feet are soft, compared with hoofs, and the spike would penetrate deeply. Unable to pull away, the hippo would then try to free itself by stepping on the log with another foot, to give itself purchase in withdrawing the first one. Thus the second foot was soon impaled on a different spike. Eventually all four feet might become engaged, whereupon the animal would lose its balance and fall down. Now the natives burst from their hideaways and finished off the hippo with their knives.

Apparently that method, effective when it worked, was not too reliable. Animals sometimes pulled free, or if they were caught by only one foot, they might drag the log into the water (where they died miserably but not to the natives' advantage). Hence the second method. This one was based on the fact that most points of emergence from rivers were along steeply sloping banks. Goodness knows it must have been difficult for the animals to get out of their bathtubs, but the return to water after a night's foraging was a breeze. Indeed the hippos often did not even bother to walk down the inclines but simply slid along them as happily as children in a public playground.

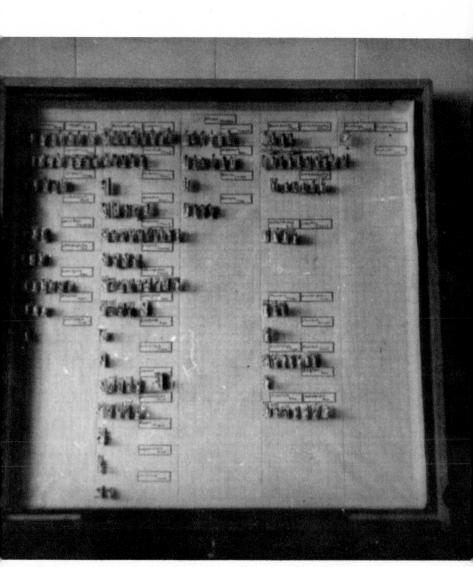

Mosquito Collection—Johannesburg Lab

The author, at Ndumu

Pottie, at Ndumu

Mamba (right) and friend, Ndumu

Mkawpis, Ndumu

Field Station at Ndumu

Bob Kokernot bleeding child at Ndumu "clinic"
Site #18, Ndumu; Giant Fig Trees

Mosquito catchers at work, Ndumu

Mosquito catchers' sisters drumming, Ndumu
Qmba Ngwenya killing mosquitoes, Ndumu

Shokwe Pan, lower end

Baobab trees, Kruger National Park

The camp at Lumbo: Mammalogy tent

Caged wild animals trapped at Lumbo and brought in for bleeding
"Lumbo," pet mongoose at Lumbo, feasting on jackal carcass

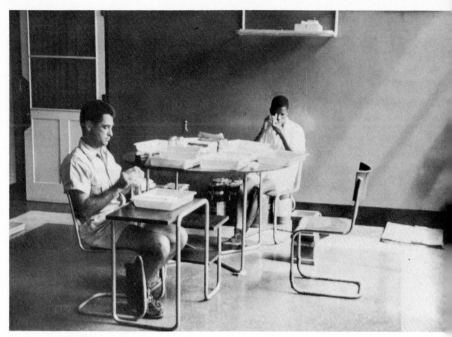

Jacinto da Sousa (L); Simão Estephão (R) identifying mosquitoes, Lumbo

Burrows of *Sesarma meinerti* crabs at Lumbo

The natives had to work fast to install killing devices at such sites. They would have prepared a different sort of log long in advance. This one was heavy also, and it had to be of extremely hard wood. One end of it was patiently sharpened to a long tapering point. The trick now was to get the log firmly into the proper position under water at the base of the slide. It must be at an angle so that the point was aimed at the oncoming animal's throat, and it must be anchored so that the impact of the behemoth would not deflect it from its intended duty as a lethal stiletto.

So far so good—one can appreciate that a game reserve at Ndumu would be a laudable innovation. But what about lions, elephants, jackals, hyenas, zebras, rhinoceroses, giraffes and various species of antelope to populate it? Can you have a proper reserve in Africa without such denizens? The answers are both yes and no. The authorities were going on the hypothesis that if the reserve were established and maintained for a sufficiently long time, some of those species would infiltrate into it. But meanwhile the only residents in the foregoing list were a few kinds of antelope, and even these were uncommon. During all the time I spent at Ndumu, I saw only three—the bushbuck and two species of duiker, red and gray—at rare intervals.

Smaller forms were present in fair abundance, but one does not normally establish a game reserve for creatures like mongooses, though even they profit from the prohibition of all hunting. A partial list of mammals at Ndumu that I can put together from several sources includes four species of nonhominid primates,* two of shrews, two of bats (though there must have been many more), ten carnivora,† the hippo, a bush pig, seven antelopes, the scaly anteater or pangolin, the aardvark, a porcupine, a squirrel, a large cane rat, nine smaller rodents and a scrub hare. The reason I

* Vervet, or green monkey; samango, or blue monkey; Cape chacma baboon; greater bush baby.
† East African civet; large spotted genet; water mongoose; white-tailed mongoose; banded mongoose; Selous' mongoose; Cape ichneumon; Cape polecat; serval cat; aardwolf.

said "nonhominid" primates is that man occupied the region in as truly an indigenous sense as the rest of the mammals, and his species therefore deserves a place on the list as assuredly as those mentioned.

Herein lay a conflict. The balance existing between man and his four-footed neighbors in the wild was such that the latter were held at low levels, though I am unable to say whether or not they were actually approaching extinction. This position was somewhat remarkable in view of the fact that in many other parts of Africa large hoofed animals coexist in great abundance with native tribes and other predators such as lions. It is the white man, of course, who is disturbing that situation. Native hunting methods are so crude that their effect on animal populations is scarcely noticeable. Since Zulus had been living at Ndumu for a long time, and white hunters have only rarely ventured there, I suspect that the hoped-for game reserve may have been badly chosen. In other words, the terrain may not have been particularly favorable for the support of great numbers of browsers and grazers.

An argument against such a conclusion can be found in the splendid condition of cattle that lived off the same terrain. Bob Kokernot told me that the local grasses, though appearing wiry and thriftless, nevertheless had exceptionally high nutritive value. Indeed, it may have been the presence of hundreds of cattle that caused antelopes to be scarce, for the vegetation could not support an indefinite number of competitors.

Bruce warned me to be very careful if I wandered through the bush away from established roads and trails. If I saw an arching tree with its crown bent to the ground, I must by all means avoid it, for this was probably a snare that had been set for bushbucks or duikers. At one time a telephone line had been set up from Ndumu to Swaziland, but natives had so persistently cut the wires to make snares that efforts to keep the line in repair had now been abandoned. Animals were supposed to walk into the wire noose, whereupon a trigger was released and the bent tree sprang to its normal position, throttling the victim.

Of course hunting by any means was now illegal, though it was commonly practiced. The authorities could see no means of getting ahead with their plans for a reserve while natives and their cattle occupied the area. The only answer was to get them out. But here there were several snags. In the first instance, there was no other place for the natives to go, since adjacent territory was already occupied by additional people and *their* cattle. Secondly, the natives had certain rights which even the apartheid-style government recognized: they could not be ejected as long as they remained law-abiding. Anyone caught in the act of setting a snare or in possession of a killed antelope was forthwith expelled to make his way elsewhere as best he could. From that standpoint the game rangers actually welcomed violators of the rules.

In this deadlocked position, the authorities resorted to insidious measures of progressive discouragement. "Let's make it unpleasant and unprofitable for natives to remain in the Reserve," they said. Accordingly various hostile edicts were issued. No new huts or structures of any kind could be built. No land not already in cultivation could be cleared to make new mealie fields. No immigrants could settle in the Reserve. Thus, if the human population increased, the excess had no choice but to move out, and it was hoped that established residents would gradually join the exodus if they did not die first.

Implementation of these regulations was often purposely harsh. I saw one old woman near the NRC camp at Ndumu laboriously clear a new field, drag out all the weeds, break up the clods and plant her maize kernels. Then she tended the resulting seedlings, cultivating them with a crude stick until they came into fine tassel. Soon the ears began to swell, and it looked as if she would have a generous crop as reward for all her hard work. One of the game rangers had observed her labors, too. But instead of putting a stop to them at the outset, he waited until the harvest was almost ready. Then he strode into the illegal plot and pulled up every last stalk.

The same man almost got himself into trouble on another occasion. There were no reverberations following his cruel treatment of the old woman. But then he burned down a church on the basis of the "no new structures" edict. True, it was a most rude kind of church, but at least it was a Christian edifice (for missionaries had been at work in Ndumu and had made a few converts). Typically, he waited until the builders had finished before setting a match to their handiwork. *That* act would have cost him his job if the story had leaked into the world press, for the parliament in Cape Town would have wished to appear contrite.

I suppose that antelopes might have increased if natives and their cattle had abandoned the Reserve. However, there was another practice (which I considered unenlightened) in which the rangers persevered, though they were not authorized to do so. In Kruger National Park, which was much larger and lay about a hundred miles north of Ndumu, some careful studies on the effects on the vegetation of periodic burning were being conducted. Until the results accrued, no one could say whether the custom was beneficial, without effect, or detrimental, and meanwhile the official policy was to hold burning in abeyance.

The natives knew—or believed that they knew—that burning grass and scrub areas was good for their cattle. Even during the dry season, a week after each fire one could see myriads of tender green blades protruding from charred wastes. But what had this to do with antelopes? Cattle were introduced species, and perhaps they needed some kind of coddling. Antelopes had existed here for as long as it took them to evolve and must be adapted to the vegetation as it exists naturally. What is good for cattle might not necessarily be good for antelopes. And fires were certainly not good for the ground, for they burned out the humus, and with the next heavy rain, erosion of the topsoil would be enormous. Indeed, exposure of the bases of grass clumps led to extensive overgrazing by cattle, thereby intensifying the potential for erosion.

Yet the rangers went about setting fires right and left. I forgot to mention that this was another new taboo for the natives—*they* were not allowed to burn their cattle ranges. Naturally the Reserve personnel fired what they considered to be antelope habitats, so that the effects did not help black Africans very much, while at the same time—in my opinion—working against the very species that were supposed to be at stage center.

Overpopulation is as great a threat to wild animals as is overhunting. Possibly illicit practices by Zulus and uninformed prejudices of game rangers were ultimately working in favor of antelopes by keeping the animals' numbers at a reasonable level, compatible with what sustenance the region afforded. But I'd like to see proof of that first, in the form of modest herds struggling to survive in the burned areas. Indeed, I'd like to see a herd in any condition. In two years at Ndumu, I never felt that I was in a game reserve at all! So-called management seemed mismanagement to me. The natives were persecuted and I studied mosquitoes: that sums it up.

Or almost. Part of my duties had to do with small rodents, which, as I have said, were David Davis's province, but since he rarely came to Ndumu, I was charged with running several trap lines and obtaining blood specimens for antibody determinations from those nongame animals.

Gertrude Stein might well have written, "A rat is a rat is a rat." That is of course true of roses also until you get to be a specialist. I don't know one rose from another, but Ndumu rats, seemingly all alike to me at first, became as distinctive as my best friends. The multimammate mouse, for example ("rat" and "mouse" can be used interchangeably in this context), was an actual pest as well as a joy in our quarters. Our laboratory benches were provided with tiers of drawers in which we kept irreplaceable records, and often I would find that a female had adopted one such drawer as a nursery, shredding the documents it contained to make a nest for her litter. Multimammate? Females had eight or more *pairs* of teats and bore correspondingly great numbers of babies.

This was, in fact, a rather ordinary looking rodent, without conspicuous markings or proportions except for its array of mammae. But what more do you want than that? Leave the stripes or unconventional configurations to the flat-chested ones. Inroads on our precious records demanded that I keep traps constantly set for rodents on the laboratory bench, an item of good housekeeping that paid off by yielding unscheduled blood specimens as a bonus. One morning I found that a Sherman box trap had been sprung— as was often the case—and I saw a tail protruding from one of its crevices. My usual procedure was to induce the animal to enter a cloth sack, in which I could manipulate it, but this tail looked a bit strange, and on examining it closely I saw that it was not sparsely haired but *scaled*. A lizard, perhaps, I thought.

I tried to peer inside the trap through various cracks but could not see the contents at all. By this time the tail had been withdrawn. Now I decided to empty the trap into a transparent plastic bag, and first thing I knew I was looking at a small snake. As I held the plastic bag up to my eyes, a spray of droplets appeared on its inner surface. A ringhals, or spitting snake! It could just as well have done that while I was gazing into the Sherman trap, and I might be blind today.

Bush rats *were* somewhat scaly-tailed, but they would not dream of entering houses, and as for the pouched mouse, the single-striped mouse, the Kalahari dormouse and the fat mouse, they, too, clung to outdoor habitats, though they exhibited individual preferences for the type of terrain they would frequent. Dormice, which were charming little furry creatures, liked Site 18 with its giant fig trees, while bush rats were most common along the Ingwavuma River at Site 26. Indeed, except for multimammate mice, which were almost everywhere, I caught more bush rats than all the less common forms combined.

At one point I almost went on strike as far as rodent collecting was concerned. David often "seconded" (assigned) one of his assistants to help me with the trapping program, and this son of an African Boer nearly drove me out of my mind. When we came in with our morning's catch, the routine required me to anesthetize each animal with an intraperitoneal injection of Nembutal and then to bleed it to death from the heart. If we had a dozen specimens or so, an enamel tray on the lab bench gradually accumulated defunct carcasses. My aide delighted in getting the next animal in readiness for my needle. Removing it from the sack, he would hold it by the tail between prongs of a large pair of forceps and then lower it so that it could run over the carcasses with its front feet.

"See what's going to happen to you next, you bastard," he would crow. Well, I am sure that the rats had no premonitions of the future, but I simply could not stand that man's attitude. On the other hand, I was a "foreigner," David was collaborating with our program by furnishing valuable assistance, and I was no physical match for the Boer even if I had tried to square things by reducing *him* to a carcass. Gad! What people this world has spawned.

The fat mouse was one of my favorites. These and dwarf mice both lived in the NRC compound, though they refrained from visiting us indoors. They were really too small to bleed, and I gratefully contented myself by letting them go after they had been trapped. Oh, I did sacrifice the first specimens in order to prepare their skins and skeletons for the record, but thereafter I was no longer their enemy. The fat mice were really *fat*. Their scientific generic name, *Steatomys*, is reflected in other words in our language, of Greek origin, such as "steatopygia," referring to excessively fatty buttocks of some Hottentot women, and "steatorrhea"—fatty stools—of sufferers from that mystifying tropical disease, sprue.

Apparently these mice are supposed to be fat. It is not a matter of their being lethargic or overindulgent. During the dry season they aestivate in burrows underground, and their accumulated adipose tissues sustain them in the same sense that obese bears are able to survive during hibernation in boreal winters.

9

AVIAN ORACLE

Rats and mice do not a game reservation make. Nor does a nose to the grindstone stimulate forever. For a change, Bruce and I took off on a jaunt to Port Elizabeth, on the coast of Cape province, to consult with Mr. Richard Liversidge, an eminent student of bird behavior, about our ornithological investigations relative to arboviruses.

It was Bruce's first air trip and he was terrified from the moment we left the ground at Johannesburg. He was careful not to reveal his feelings until afterwards, but when we finally set foot on the ground, he commented on the uniform fortitude of the rest of the passengers. We had gone through a squall near East London, and the plane hit some hard bumps. "But they all kept on leafing through *Life* and *Reader's Digest* as if they must finish before the crash," he said. "I never found those periodicals *that* interesting."

Richard Liversidge and Dr. G. R. McLachlan, of the Port Elizabeth Museum, had recently completed a revision of Roberts' *Birds of South Africa*, so Bruce and I had come with our questions to headquarters, as it were. Not that many other fine ornithologists did not exist in South Africa, but Port Elizabeth had become an important center of information, particularly because all the bird-ringing data were sent here for processing and analysis. The latter encompassed Bruce's chief interest, for the movements of some birds over wide areas might account for dispersal of so-called avian viruses for corresponding distances, and the documented migrations of known, marked individual birds provided exact information that could not be gleaned simply from knowledge of a species' general range.

We were welcomed and taken to our hotel by Richard, a slender young man with thinly horn-rimmed glasses, who shortly thereafter introduced us to his senior colleague in their joint professional office. Our hosts occupied an institution in Port Elizabeth such as I have not seen elsewhere. This was a combined museum *and* zoo. The idea is a splendid one, for ordinary natural history museums, no matter how outstanding their presentations of mounted specimens and painted habitat backgrounds, must build solely on the fact of death. But here one could prowl through aisles of glassed-in effigies and then emerge into sunlight to see many of the same reptiles or birds alive and in action.

Unfortunately, bird-banding records were in a mess. I cannot remember all the reasons why, but at least two of them were that Dr. McLachlan had been sick recently and that a new system of filing had just been inaugurated. Perhaps the records had also been moved from one repository to another. Anyhow, the cards were now stacked in shoeboxes and other miscellaneous containers on top of filing cabinets, bookshelves, odd tables and so forth and could not possibly be examined by us until everything had been straightened out—a task that would require not less than a year.

Under those circumstances Bruce and I might as well have remained in Johannesburg. But ornithologists are a cordial lot among themselves. As long as we had arrived in Port Elizabeth (safely!), McLachlan and Liversidge now collaborated on our entertainment as effectively as they had revised Roberts' book as a team. It turned out that Dr. McLachlan owned a cabin on a small stream near the Addo Game Reserve and that Richard was con-

ducting bird censuses at several specific loci in that region.
Wouldn't Bruce and I like to take a field trip with Richard and
enjoy Dr. McLachlan's hospitality in his absence? Here—take Mc-
Lachlan's key to the cabin. Tinned food was on the shelves, and a
bottle of brandy could be found if one rummaged thirstily
enough.

That was too good to miss. Bruce and I accepted immediately,
and next morning Richard picked us up at the hotel in a Land
Rover. My immediate timid question was whether there was any
place we could go along the coast to see either a penguin or an
albatross—or even both. I knew from Roberts (or, as I should
now say, McLachlan and Liversidge) that the Jackass Penguin
bred on offshore islands all around the southern tip of Africa, and
I had never met a wild member of that family. Albatrosses would
be equal novelties. Richard answered that to see both birds would
require a special excursion by boat, and besides, the studies he
was making were of terrestrial species. Naturally I could not divert him from his projects: Bruce and I were unexpected accretions
in the Land Rover, and our host was already displaying hospitality beyond the conventional boundary. I tucked penguins and albatrosses back into my subconscious, whence I could summon
them at a more auspicious time.

(After our return to Joburg, I received a letter from Richard,
saying that a Jackass Penguin had been found in the bay on the
very day we were there. A fisherman had picked it up and
brought it to the wharf—which we had visited a half hour previously! It had become saturated with waste oil from a ship and was
now incapacitated. That bird *would* have counted on my life-list.
I have run across several derelicts in that unfortunate state on
beaches of various continents, but would prefer not to have them
on the list if oiling means the life of each one—as it almost surely
does.)

Richard had recently returned from a year of graduate training
in England and was imbued with the latest philosophies of bird
study, which is to say that you scarcely dare get near your subjects

lest your presence cause an alteration in their normal behavior. His slight frame seemed to fit him for such easy-going methods, but I soon learned that he approached other matters vigorously enough. In the first place, he drove the Land Rover with confidence, mastery and arrogance. Then, when we had stopped at a tidal estuary to look at shore birds on some exposed mud flats, he suddenly spotted—of all things—a sheep far out in the water.

"She's got a drowned lamb out there," he said while peering through his binoculars. "The lamb has drifted farther than the ewe can go, but she is staying as close as she can."

Off came Richard's clothes, except for his shorts, and soon he returned with the sopping ewe in his arms. He set her on the ground and shooed her into the bush.

"Sorry," he said. "Let's get on."

Dr. McLachlan's cabin was really nothing more than a crude wooden structure, but that is exactly what it should have been. The stream came down through a narrow valley between lofty and precipitous hills on each side, at the site of the cabin widening out into a deep, almost circular pool. In fact, we had to leave the Land Rover some distance below and wade in order to reach the camp. The solitude and the grandeur of the scene would have been ruined by a pretentious cabin. As it was, the small unpainted building made of rough boards almost disappeared into the background, and one felt as if man's contamination had not yet reached this primeval recess.

Bruce was so moved that he immediately became primeval himself, shedding every man-made stitch and arching into the pool from a large rock. Richard, although having been largely immersed along with the sheep only an hour ago, was quick to follow—minus shorts this time. The water had been far too cold for my taste as we waded upstream, so I contented myself by watching my companions for a while and then went in to search for the brandy bottle in the place that Dr. McLachlan said it ought to be. It was. And the swimmers soon found that they, too, needed the warming ingredient it contained.

As we sat on the rock eating our sandwiches (for we did not disturb the canned goods), Richard pointed to some tall plants, somewhat resembling small palms, on the slopes above us. He said they were cycads, which are the most primitive of seed-bearing plants. An important source of coal in past ages, they are now only rare remnants in the vegetable kingdom—living fossils, as they say—and to see them growing wild was a botanical treat. Unfortunately they were all too high for us to climb for a better look, but through binoculars I could imagine that I saw their struggle for survival. The hillsides were too barren and rocky for many other plants to find it worthwhile to colonize them, and the cycads appeared to be outcasts whose only hope was to accept an environment that was not in demand by stronger forms.

On the way to this place we had passed through an extensive zone where an entirely different sort of vegetational war was in progress. Many years ago someone had imported specimens of our southwestern prickly-pear cactus. I believe that the purpose origi-nally in mind was to make hedges of them for fencing cattle. But as sometimes happens—as with our English Sparrows and Euro-pean Starlings—the "foreigner" unexpectedly made itself too much at home. The cacti were not content to confine themselves to border duties. Slowly at first, and later rapidly, they spread out on both sides. Eventually they formed solid masses covering many square miles, and all other vegetation was choked out. Now the people of the region were attacking the cacti with bulldozers. We saw truckloads of the thorny plants being taken to a crushing fac-tory where they would be converted to pulp that might be used as fertilizer or fodder (though I wondered whether those spines had been sufficiently eliminated). But it was still questionable whether the cactus could be stopped and the cleared ground re-stored to its former uses.

Although this was a bird trip, we were alert to anything else that came along. When Richard pointed to some animals about the size of rabbits high on the rocks above us and said that they were dassies, I did not exclaim, as I should have, for the name meant nothing in my vocabulary. Then, however, he mentioned two synonyms, "Cape hyrax" and "cony," and comprehension led to enthusiasm. This was another evolutionary remnant—or perhaps "oddity" is a better word—though apparently getting along better today than cycads do.

Zoologists don't know exactly where to place hyraxes in relation to other mammals. Their teeth have some similarities to those of horses and rhinoceri, and they might have a distant ancestral connection with those groups, unlikely though that would seem. Of course, we could not get a close view of them, but my books tell me that the forefeet have four short toes *without* claws, and the rear ones have only three toes, each with a flattened nail. No other mammal has appendages anything like that. Dassies are found only in Africa and Syria, being mentioned in the Bible as "conies."

All these diversions were great, but now it was time for Richard to visit his study areas. We cleared up our sandwich papers, restored the brandy bottle (now containing no more than a few drops) to its hiding place and locked the cabin door. This was going to be a fine show, for I had never seen a British-trained ornithologist in action in the field.

The first locus was within walking (not wading) distance of the pool. Upstream the valley broadened out slightly and a grove of moderately tall trees occupied a level area on one side. Before we reached that spot, Richard stopped us to make the following pronouncement:

"My procedure is to sit on a certain log for half an hour in complete silence and immobility except for the use of my pencil. Whenever I hear or see a bird, I tick it off according to the time, the direction, the distance and the species, as well as its age and its sex. Whatever I can tell about behavior, feeding, or change of position of an individual bird is recorded also. I do not even turn my head to see a bird if it is to one side; my ears have to inform me what kind it is and what it is doing. My chief design is to be so inconspicuous that the birds go about their ways as if I were not there at all. Naturally that is an impossibility, but I approach the ideal as closely as my arrested presence will permit. Therefore, if a mosquito bites me, I do not slap it. If my nose itches, I do not scratch. If I become uncomfortable, I do not shift my position.

"It is essential that you do the same. Moreover, I do not want you to sit near me, because the birds might become alarmed on seeing a crowd. No smoking, of course. I shall tell you when the time is up. Later I can analyze my notes and compare the abundance and activities of birds occupying this habitat with data from former visits. That gives me a dynamic record of changes taking place in the avifauna in relation to local and long-distance migrations, seasonal differences in food supply, territorial competition when nests are being attended, and so on. Now please be seated, gentlemen—you over there, Bruce, and Brooke here."

Thereupon I underwent one of the most agonizing half-hours of my life, and Bruce agreed afterwards that he had hardly been able to endure it either. If only Richard had not mentioned itchy noses, perhaps mine would not have acted the way it did. I was reminded of my boyhood, when I had to sit in silence through Quaker Meeting, but at least you could scratch politely there, particularly if Mother wasn't looking (and she *was* supposed to be contemplating). Moreover, my slight deafness prevented me from hearing birds that Richard was "ticking off," and if I had heard them, I would not have known what they were. Even Bruce, whose ears were keen, failed to recognize most of the calls, for the avifauna was markedly different in the southern part of Cape province from that which he knew well in Transvaal.

By the time we had waded back to the Land Rover and driven to Richard's next two study areas, going through the same mute, immobilized ritual at each one, Bruce and I had achieved a high respect for modern ornithological methods, concluding at the same time that we could not aspire to proficiency as practitioners of those arts. My lord! You have to be a martyr of some sort to do that kind of work. I love birds and think I know something about them, but Richard's performance threw that conviction into perilous doubt. If it is not valid to stand up, move about, and even *chase* birds when that must be done to get them within range of binoculars, the only logical consequence to entertain is that I have *never* seen a bird behaving normally. In other words, I can know nothing about them.

That was a chastening lesson. However, Richard was by no means dedicated exclusively to his own studies. That evening he had arranged for us to accompany a colleague from the museum who was engaged in making tape recordings of bird voices. After dinner we drove to a beach. Port Elizabeth is on the southern margin of the Cape, essentially at the same degree south as Cape Town, and on the maps that I can find this could be the South Atlantic, the Indian, or even the Antarctic Ocean, or perhaps the junction of all three. Whatever its name, this shore had been a strategic one during World War II. Just before the Japanese seized the East Indies, but when it was clear that they were going to, Allied ships made a run for western ports with as much raw latex as they could load. The Red Sea, Suez and Gibraltar were fairly well blockaded, so captains of the vessels chose to hazard a course around the tip of South Africa. German submarines were prepared for that maneuver and sank many of the freighters. Now, almost two decades later, deteriorating hulks at great depths were still yielding bales of latex sheets that washed ashore from time to time, the crude material largely unaltered physically and chemically by its long submersion. Many beachcombers were still making what they called a living by salvaging those once-treasured flabby slabs. Actually, the sheets may no longer have been susceptible to processing into fine rubber, but they were at least temporarily serviceable in their present state. Richard had lined the floor of his Land Rover with them and his friend, whom we now met, had done likewise.

The voice-recording apparatus was as sophisticated in its way as Richard's bird-watching techniques had been. An entire truck had been equipped for the enterprise, the outstanding gadget being a huge parabolic receiver mounted on a tripod. To back up that basic appurtenance, a host of electronic accessories lurked behind the scenes in the bowels of the truck, the various circuits ending in a simple box designed to take down incoming decibels in wax or on tape.

That night was devoted to immortalizing the babble of a flock of Sandwich Terns roosting on a sandbar about a quarter of a mile distant. These were "wintering" birds, in the sense that it was now summer in the southern hemisphere and the terns had fled from the northern winter. In their hundreds they behaved like so many other vacationers, never ceasing in their conversation. A combination of sussurating waves, the soughing of a rising wind that presaged rain tomorrow and the tern voices made a memorable tape for the ornithologist as well as an impression—electronic in its own physiological way—on my memory that was equally durable.

The morning lived up to auguries of the previous evening. When Richard called for us at the hotel, a heavy drizzle or light rain was falling.

"Have you ever seen Cape Sugarbirds?" asked Richard.

Well! Who cared about rain now? Here was a last-ditch opportunity for a new avian family on my list. In my quest for the more spectacular penguins and albatrosses I had forgotten all about sugarbirds. Anyhow, I would have thought that another special trip to see them must be made. But no. Richard said that he knew of a hillside not far away that was being ravaged for a suburban housing development, but that the natural vegetation had not yet been fully destroyed, and a few sugarbirds were still in residence among the protea bushes to which they are almost exclusively adapted. The protea is remarkable in being confined to the southern hemisphere; in addition it is a national emblem of South Africa. Its dried inflorescences make handsome permanent indoor decorations in vases. That a unique kind of bird should form binding associations with a unique kind of vegetable is itself unique. The case may go farther than that, for sugarbirds feed not merely on nectar from protea flowers but also on attending insects in the blossoms. Indeed their diet is largely insectivorous, so that one wonders whether the protea insects may not also be confined to those plants.

Honey-eaters, among which sugarbirds are included by some taxonomists, all occur in Australia and the South Pacific with the exception of the two South African species. Thus I would have to go even farther afield than was necessary for seeing penguins and albatrosses, should I miss my sugarbird here. Continuing his role as the perfect host, Richard donned his raincoat and led us to the bulldozed area. Bruce and I sloshed through the mud. Our binoculars became useless almost at once, for when we wiped the lenses they became smeared, and after that they seemed to invite haze. Consequently we resorted to the primitive naked-eye type of observation. The sky was not giving us any help, for its lack of directly transmitted light rays rendered all objects flat and incorporeal. The landscape might as well have been a paper cutout.

Yet the proteas were *there*. Nothing could alter that fact. Indeed, I was almost content from that standpoint alone. But "almost" is not the same as "quite." Richard was apologetically prefacing a remark about our giving up the search when I exclaimed, "There's one!" That might well have been the result of overeagerness. If you want to see flying saucers, it is easy to imagine that you do, when the reality is only a puff of smoke. Richard's confirmation therefore came as a warming statement.

"Yes," he said, "a fine male, too. A bit bedraggled in this weather, and the long tailfeathers are drooping heavily. Say, I wonder how they manage to dry out? Perhaps if we sat here and . . ."

But now Bruce and I were off to Johannesburg, squalls or not. After all, he was by this time a blithe veteran of the air.

10

WINTER IN JOBURG

Although ABVRU's field program was centered at Ndumu, the lab in Joburg sometimes received information about arbovirus activity elsewhere, and no one was at all averse to looking at mosquitoes, farm animals, people or any other sort of specimen from the nontropical Highveld plateau. In fact, the year before I arrived in South Africa, ABVRU had sent an expedition to Cape province to investigate a disease of sheep. Here they found mosquitoes biting lambs at a place called Middelburg, and when the cause of the malady was eventually isolated and found to be new, they inevitably named it Middelburg virus.

Not only in South Africa but almost everywhere else that arboviruses have been studied, scientists have noted that a given virus often has years of special prevalence, such as our confrontations with Asian and other brands of influenza from time to time. Middelburg showed itself to be that kind of actor by being discovered first in Cape province in 1957 but then putting in a second appearance at Ndumu in the same year. Ken Smithburn's original Simbu Pan Six thereby promptly became a Tongaland Septet.

Thus the lab remained constantly ready to answer calls for the investigation of unexplained epidemics in human populations or epizootics in subhuman ones. Naturally, most alerts of that sort were sounded in warm weather when mosquitoes were on the wing. But winter in Joburg remained as busy a time as ever for all of us at ABVRU. There were endless reports to write to New York headquarters, while in the various lab rooms innumerable baby mice were inoculated with the accumulated backlog of last summer's frozen mosquitoes. Even new isolations of familiar viruses could not be positively identified until several different tests had been completed. A really new virus—that is, one hitherto unknown—might require more than a year's work before it could be characterized sufficiently to prove that it *was* different from any other kind.

Since I spent most of the hot months at Ndumu, I now invariably think of Johannesburg as the cold place I came to know in winter. The center of the city was extremely modern, with buildings they called skyscrapers, and those which soon could truthfully claim that term at the rate each new edifice was overtopping the last. I went into town as seldom as possible, for I did not like this city any better than all the other cities I have had to face. However, there was one feature of Joburg that made it different and indeed interesting, if not attractive, and that was the mine dumps. At the least expected places you would come to a dead-end street and see a towering mound of light-yellow sand soaring even higher than the buildings. From an airplane it was evident that the city had been built around and between a host of such artificial hummocks. These were more ancient than the metropolis, for they used to lie in isolation when Joburg was no more than a crossroads in the wilderness seventy years ago. Of course they were only diminutive hills then, but as the quest for gold ran deeper, they rose proportionately.

One could make revealing studies of those mounds if one had the patience. In a real sense they were aboriginal, their substance not having been touched by sunlight since this planet congealed. Therefore the hand of life, fingering triturated rock fragments that had come from a mile below present department stores and booteries, must have paralleled phenomena that attended the first essays of lichens to colonize raw rock surfaces when plants managed to leave the sea. Much has been said of the cataclysmic explosion of Krakatoa, off the coast of Sumatra, and the sequence of its subsequent repopulation by vegetation and animal life. The same can be documented on more recent volcanic slopes. But most of those places are distant and difficult of access. Mine dumps in Johannesburg put that study on the doorstep of a fine university, and I hope some young professors are taking their classes on field trips to observe the glacier-slow changes that attend new biotic communities in the making.

In the beginning these mounds are a hostile environment, both acidic and unstable. Rain comes to a mixed sort of rescue, leaching out excessively concentrated chemical compounds but also eroding the piles, so that whatever began to take a foothold at higher levels was undermined, while colonists near bases were interred. Weathering, leaching and erosion ultimately resulted in the formation of furrowed hills that had lost their pristine yellow sheen and now looked like dull gray sterile wastelands. However, their very sterility rendered them now more benign than they had been previously. Almost unbelievably the stabilized slopes began to clothe themselves with grasses and saplings. Given enough time, they will be Johannesburg's parks. Unless—oh, good heavens!—somebody discovered that detritus from the mines, freed of its gold content and piled up for decades, contains uranium in quantities worth salvaging. Since the work of bringing all those tons of material has already been laboriously done, all that is necessary today is to go through those hills all over again. After that, perhaps biology can have its undisturbed way.

The natural history of man is tied to mine dumps in a variety of ways. Supreme predators, the tycoons, are at the head of the ecological food chain, while NRC camps, such as ours at Ndumu, represent bacteria and amebas at the lowest level. The tycoons are aware of their heart attacks while the amebas don't know what is going on at any time, so you can decide which you would rather be. Nevertheless I used to be ashamed, in a way that I still don't understand, when we attended shows put on by ignorant "amebas" and saw all the still hale tycoons congratulating themselves.

These spectacles were called mine dances, and they were really remarkable, for each tribe had its own traditionally set dance which could differ outstandingly from one regional group to the next. Some were slow and deliberate, as if portraying a stealthy approach to a wild animal. Others were frankly sexual, with rapid undulations of the body. The rest were more difficult to identify, though fighting was easy to spot in parts of some of them.

Usually the dances were accompanied by singing and shouting. In addition, the natives often wore metallic discs which jingled in time to their body rhythm. In the foreground there might be non-performing spectators—members of the same tribe—who chanted, clapped their hands, or played an accompaniment on drums, on sheets of corrugated iron roofing, or on a strange kind of marimbalike instrument that might be made from hollow pieces of wood or from old kerosene tins and oil barrels.

The dances were held on Sunday mornings and were well attended. If you had overseas guests you would not dream of having them miss this display. Admission was charged and tea and biscuits could be bought during the mid-morning break. Printed programs outlined the sequence of acts, giving the name of each tribe, its source of origin, and a few words about the significance of individual dances. A public-address system was utilized to announce program changes and also to indicate the judges' decisions about prizes at the end of the show.

All gate receipts and revenues from the tea counter went into a fund for the benefit—not of gold magnates—but of the natives. This sounded unexpectedly benevolent to me when I first heard about it, but its logic eventually became apparent. Just as a good carpenter takes great care of his tools, so the cheap laborer had to be nurtured assiduously for the mine to be profitable. These tribesmen were here without their wives (if they had any) and were housed on an ethnic basis in dormitories within each mine compound—Xhosas in one, Zulus in another, Chizwinas in a third, and so on. This kind of distribution was highly important for their morale, since a mixed group, each member coming from a different tribe, might be incapable of holding a conversation. Mine-dance competitions were further morale-builders, taking the place of sports. During the week, when not working underground, the natives rehearsed for next Sunday's show during leisure hours.

A very fine medical service was maintained to protect the workers' health. Physical accidents were unavoidable, of course, but a particular occupational hazard in drilling ore at great depths was dust inhalation that sometimes led to a fibrotic condition of the lungs called pneumoconiosis. That, in turn, might promote tuberculosis. Paul Weinbren's father was chief radiologist to the mining syndicate and must have looked at thousands of X rays of hardened or cavitous lungs.

During his year at the mines, perhaps many hundreds of miles away from his home in some primitive part of the Union or even in Moçambique or other foreign places, a native was restricted to the mine compound. He was not allowed to wander around Joburg, so that when he left the city virtually his only knowledge was of the deep shafts in which he had worked. As an educational experience his year was totally void. One might wonder what induced him to come here in the first place, giving up a year of his life (or two years, if he cared to contract for an additional tour). He was not attached to the white man, and besides he might be endangering his longevity.

The answer was money. The miners had all seen some of their predecessors returning to their kraals after a year's absence, now suddenly able to buy enough cattle to exchange them for a wife—maybe even two wives. That incentive was more than enough to induce them to accept the hard labor and the social privations entailed.

Yet some of them went home almost penniless. They probably weren't paid much anyhow, but at the end of their terms, they were allowed a carefully guarded shopping spree in Johannesburg. Not knowing much about money, they were shockingly exploited in the cheap stores to which they were taken. Another thing their predecessors had done to arouse envy was to arrive back at their villages with gaily painted foot lockers containing such wonders as bright-colored blankets, mouth organs, pennywhistles and other assorted junk. No doubt they had paid several times the true value for those articles (provided they *had* any appreciable value—for the foot lockers were so flimsy that their paint was the only conceivable selling point; the blankets were thin and desirable chiefly for their tints; and junk is junk anywhere, especially when it breaks down after a few days).

I wondered why miners had to be recruited in the first place. It placed an ultimate hardship on everybody. Wouldn't a permanent working force be more economical—and reasonable? The answer to that proposition is that men would not come to Johannesburg permanently without women, and each man allowed to reside there must consequently be thought of as representing a populous household. City authorities said there wasn't room for that many additional people, though it was equally true that no one wanted them. It was far more satisfactory, despite all the trouble involved, to turn over the labor force once every year as long as the supply of bucolic natives lasted. And there seemed to be no bottom to that barrel.

The recruiting system can be compared with a carefully planned game conservation program in which the best breeding stock is manipulated in a way to promote reproduction among its outstanding members. One can see that recognition of certain rights among natives in Ndumu was not fully altruistic. The forces of the mining syndicate were strong at high government levels. Antelopes be damned! Let's breed blacks instead, so that they can come to work in the mines and entertain us on Sundays; and let surviving studs go back to the Usutu River burdened with trinkets and bulging testicles.

Most of our visitors came during winter, i.e., when they were having their summer vacations in the northern hemisphere. They were often unprepared for Joburg's climate, and I saw many of them shivering and drinking extra cups of hot tea at the mine dances. This was the dry season. Clear cold mornings were frequently accompanied by heavy frosts. Lawns and garden flowers became brown for more than lack of water, and bird baths at fancy residences in Rosebank were filmed over with ice, if not frozen solid. That is what Miami would be like if it were a mile high.

Snow was a rarity, relatively speaking, though I did see a couple of falls that approached an inch in depth. It usually melted the same day. There was one occasion, early in the history of this city, when a real blizzard struck, and downtown Joburg was buried almost two feet deep. Nobody knew what to do except the news photographers, who made incontestable pictures of a fact that might otherwise have become magnified in legend.

Of course, deciduous trees lost their leaves. The scene, then, often resembled one such as might be viewed in the temperate zone on a January day, and that did not please me at all. Field studies at Ndumu were largely canceled for the time being, because mosquitoes became scarce at that season—not so much from the cold as that their aquatic breeding places dried up. Consequently we had very few recoveries of viruses during winter. At first I accepted that lacuna as one of the facts of nature, but on reflection I could not bring myself to be quite so easily convinced. If viruses were going to reappear next summer, they must have survived somewhere, and since their period of circulation in the peripheral blood of infected vertebrates is so brief, they must have wintered over in resting mosquitoes. The trick would be to find where such mosquitoes hid themselves. Our mosquito catchers must be taught new techniques for finding them. If necessary, we must get additional mosquito catchers. In short, we were faced with the need to make extra efforts in winter in order to explain the phenomena of summer.

At least that was how I liked to look at it—anything to get away from Joburg's bleak aspect. Of course I was not so wise during my first winter there. I was still becoming oriented, and in the sort of work ABVRU was doing, this takes the better part of a year. Pottie and I did make one or two sallies to the field station, but that was before I had taught mosquito catchers to canvass tree cavities, rock crevices and ground holes, and we drew a virus blank according to expectation. I am not implying that the following winter solved every secret, but we did then capture lots more insects and the lab rejoiced in an occasional virus isolation. The principle of viral over-wintering in mosquitoes was probably correct.

11

GERMISTON LAKE

A somewhat wintry mosquito-catching chore that fell to me in the Highveld was discharged during warmer times of year, but in the late afternoons and evenings, when it usually became distinctly chilly. Earlier I referred to the dreary industrial town of Germiston, near Joburg, where we used to get our dry ice, and now I must refer to its golf course.

Just before I came to Johannesburg, ABVRU had been through a medical-detective game such as Berton Roueché writes about, except that he always gets his "man," while ABVRU, though indeed catching two criminals, got the wrong ones. A recent epidemic of febrile illness among suburban Germiston residents living near the golf course had baffled local physicians as to its etiology. Clinical signs were somewhat suggestive of arbovirus infection, so our lab was called upon to make investigations. That required, of course, the collection of many blood specimens from patients and their contacts, but in addition a survey of local mosquitoes was strongly in order.

It took no more than a glance to disclose the obvious nearby place where mosquitoes might be breeding. Patients' residences were adjacent to a fair-sized lake, about a mile long and half a mile wide. This was probably an artificial lagoon, owing much of its content to underground water pumped up at adjacent mines and discharged into the lake, for there were two large mine dumps a short distance beyond its northern border. However, as in other cases that I have mentioned, a high content of dissolved chemicals was not incompatible with the existence of some forms of life, and the lake was rimmed for about a third of its perimeter by marginal reed beds. Artificial or not, and mine dumps in the background or not, the watery expanse was scenic, and it was natural that homes should be built near it, and also that a park and a golf course should have been sited at other places along its fringe. Suburban residents probably saved mosquitoes the trouble of searching them out in their homes by walking in the park in the evenings or sipping cocktails on the club's terrace.

At the far side of the lake from the homes, the reeds grew thickest, extending fifty yards from shore at one point. This was directly in front of the clubhouse. Here great numbers of birds congregated to spend the night. Club members, enjoying their "nineteenth hole," could watch Cattle Egrets, Sacred Ibises, Blackheaded Herons, Masked Weavers, Pied Starlings and Laughing Doves coming in to roost, while Moorhens, Redknobbed Coots and a species of Old-world Warbler that I failed to identify further were resident. Cattle Egrets actually nested there in abundance, though they had to forage at a distance for food for their fledglings and could therefore be regarded as commuting residents as truly as the human ones.

Personnel from ABVRU at once made contact with management of the golf club and obtained permission to invade the premises whenever they wished. That did not conflict greatly with golfers' pastimes, for mosquito collecting was undertaken chiefly at dusk. At least nobody got hit by golf balls and no complaints were lodged by players when ABVRU lab attendants, pressed into duty as mosquito catchers, stationed themselves on fairways and greens to intercept biting insects.

Relatively few mosquitoes were caught. They could be reckoned only in hundreds, rather than in thousands or tens of thousands as at Ndumu. Yet three strains of virus were isolated from them—one from a mixed suspension containing five *Culex* (*Culex*) *theileri* and twelve *Culex* (*Neoculex*) *rubinotus;* and the other two from pure suspensions of eleven and twenty *C. rubinotus* respectively. The names in parentheses denote different subgenera of the genus *Culex*, indicating that *C. theileri* and *C. rubinotus* are only distant relatives within their genus. Thus one would suspect that even in the mixed suspension, *C. rubinotus* had been the virus carrier and *C. theileri* had been innocent. *Culex rubinotus* made its debut as a virus host on that occasion. The viruses, too, were new ones, the first two strains being alike and falling into a known group but possessing some unique serological characteristics. These were promptly and appropriately named Germiston virus. The third one showed no affinities to any known viruses. It was called Witwatersrand virus in relation to the gold-bearing strip of the region.

But much to everyone's astonishment, sera from febrile patients at the opposite end of the lake did not react with either type of virus. Medical detection broke down in this instance. One might then jump to an assumption that mosquitoes had become infected with viruses adapted exclusively to birds in the reed bed, except that two of our laboratory personnel acquired indubitable Germiston virus infections while working with that agent, so that it *can* affect man and need not necessarily be confined to birds. Probably birds *were* the source, and the virus is simply versatile or not choosy about the sorts of warm-blooded vertebrates in which it will prosper. (The illnesses were mild in one of our cases and only moderately severe in the other. Both individuals made complete recoveries.)

Culex rubinotus is a rather uncommon mosquito and not much is known of its life history. Following the isolation of viruses on the golf course, ABVRU sent mosquito catchers to make further collections, but *C. rubinotus* had disappeared. Possibly this kind of mosquito was present in appreciable numbers only at one brief period during the year. If so, it would make sense to return to

Germiston Lake around the same dates as when the original collections had been made. Accordingly, I was first filled in with the history I have just recited and then directed to several red circles enclosing numerals on a calendar.

Pottie and I drove over one afternoon to find whether it would still be acceptable for us to make use of the golf course as a mosquito-collecting area. The manager, to whom Pottie introduced me, was extremely cordial and said that we could suit our pleasure at all times. I am certain that he had no inkling of *what* we were doing, but the *why* of it was clear enough: we were from the virus laboratory and some sort of public-health menace lurked in the reed beds or their denizens. As investigators, we were obviously his allies, and he offered to render extra help if we would but tell him our needs. He was disappointed to be informed that we were self-sufficient, so he simply created his own method for showing appreciation. When the sun lowered in the west and the air began to feel shivery, he would henceforth send a white-jacketed waiter across the fairway to inquire whether we preferred martinis or Manhattans this time. Subsequently the servant returned with a small table covered with a cloth, a generously filled pitcher, canapés, glasses, serviettes, etc. This was mosquito collecting at its acme.

But the upshot of this return venture, following the epidemic of a year ago, was that there were now virtually no mosquitoes in the area. The few that were taken did not include *C. rubinotus*, being chiefly *C. theileri*. However, there remained a chance that *C. rubinotus* was simply being temporarily evasive. Perhaps its season would be later this year. With that possibility in mind, we lowered our sights to the earlier stages—larvae and pupae—that could be dipped from weedy edges of a small stream that emptied into the lake. These we took to the lab, in hopes of eventually establishing captive mosquito colonies. Many larvae died, though some larger ones pupated successfully. Soon we were looking at emerged adults. Practically all of them were *C. theileri*, and there was not a single *C. rubinotus* among them. We were glad enough to have *C. theileri* just the same, for though it had been the less likely vector in the discovery of Germiston virus, the species had been incriminated elsewhere in the Highveld during an outbreak of Rift Valley fever among livestock. Consequently we returned several other times to the golf course. We never did find *C. rubinotus,* but *C. theileri* continued to appear in our dippers.

An insectary in which colonies of mosquitoes are maintained is an important adjunct to a virus lab from several standpoints. The most obvious one involves the capability of given kinds of mosquito actually to transmit specific kinds of viruses. For example, although *C. rubinotus* was proved to contain Germiston virus when captured, no one could state that this insect was able to inject it when taking a blood meal. Perhaps the virus was locked in its body cavity, unable to station itself in the salivary glands. In such a contingency the mosquito could bite warm-blooded hosts repeatedly without transferring the infection. The answer to that question could not be revealed by field studies. One now wanted recently emerged insectary specimens that would bite inoculated animals during their viremic stage. After a suitable lapse of time for viral multiplication to take place, the mosquitoes could be offered fresh nonviremic hosts for a second feeding. If these animals developed viremia, one would have convincing evidence that the mosquito had succeeded in transferring virus from one host to another.

There are many tricks in getting colonies started. But most of the tricks are not yet known, and the number of attempts that have succeeded is very small compared with the many trials that have been made. For some reason household mosquitoes are more easily colonized than "wilder" species. Among such bedroom pests, *Culex pipiens* and at least one of its variants, as well as *Aedes aegypti,* have long been standard laboratory animals in the study of mosquito-borne diseases of all sorts, in addition to serving in tests of insecticides. *Anopheles quadrimaculatus,* the former vector of malaria in our southeast before the scourge was wiped out, has been successfully colonized also; that species, too, enters houses freely.

I am not aware that *C. theileri* comes into the house-frequenting clique. At least it did not act as if that were its habit, for I can aver that it resisted becoming colonized in every conceivable way. The hours I spent nursemaiding those insects! Let's begin with the mosquitoes that dutifully emerged from pupae that we had collected. Mosquitoes' first aim in life is to mate, for they have no time to lose. Of a hundred female mosquitoes that rise from their natal ponds and streams, only a small fraction may live to bite a vertebrate animal, digest its blood, ripen eggs, and succeed in returning to a suitable place for their deposition. In some instances the entire hundred females may be lost before that simple cycle, requiring only ten days or less, has been completed.

The mating urge of mosquitoes is a strong one, as in almost all forms of life. Granted that the urge can be suspended in some animals, for example in bitches between their periods of heat, mosquitoes cannot claim delays but must get on with it as promptly as possible. To ensure a meeting of the opposite sexes, certain signals are used. Sometimes they take the form of a swarm of males flying about in a dense, stationary column and making quite a bit of humming in doing so. Females recognize a swarm of their own species by the type of aerial display and possibly also by its sound. They dart into it and are immediately seized and fertilized.

Mating antics characterized by other types of stereotyped behavior have been described for numerous sorts of mosquito. The trouble with insectaries, then, is that one usually does not know what environmental props to provide for a species, heretofore undomesticated, to accept. Simply allowing pupae to yield adult male and female specimens in the same cage is far from the requisite need for mating to take place. In other words, the mosquitoes do not recognize each other as such: it is behavior that guides them into conjugal unions.

Then there are physical factors such as light, heat and humidity that may have a bearing on mating. In general, evening twilight is most popular. Therefore one fiddles about with illumination in the insectary, using a dim bulb and perhaps putting a transparent blue shield around it. Heat and humidity can be regulated by expensive electronic controls. Space is often another critical factor, for if males get into the mood for swarming (which is what their instincts dictate they should do), they may be prevented from consummating their joint aerial performance by being housed into too small a cage. Finally, what about numbers? Despite our efforts to collect as many pupae as possible, there were rarely more than a score of emerged males in a cage. Usually no more than ten or twelve were present. Can such small numbers of individuals be called swarms?

We hoped to keep adults of both sexes alive for a considerable period so that, as males continued to emerge day by day, the colony would build itself up to swarming strength. Many a night I went to the insectary and sat by the dull light of a muffled flashlight to see whether the latest crop had overcome mating reluctance. The trouble was that earlier males had died, and we never passed the swarming threshold—provided that *was* our problem. During my vigils the males simply perched and did nothing.

Deaths among larvae were common and appeared to result from fungus infections. Therefore we specialized more and more in the collection of pupae. If we could just get a few females to mate and oviposit, progeny from their eggs might be more tractable, accepting the insectary's environment as a "normal" one, since they had been born in it. But the females obtained from wild pupae would not even bite, so they were bound not to lay eggs. As I sat in semidarkness with the flashlight in one hand, I had the other extended through a cloth sleeve into the cage, hoping to feel the sting of a proboscis. Some of the mosquitoes alit on my proffered appendage, but only as if it were an inanimate fixture unrelated to appetites.

Perhaps they would like something else. I tried naked, pink, helpless, two-day-old white mice, but the mosquitoes ignored those too. Nor would they bite adult mice. Many species of *Culex* are known to be partial to birds. Hadn't the parents of those specimens probably fed on herons in the reed bed? I therefore left a baby chick in the cage overnight, tying it up in the toe of an old nylon stocking so that it could not move but a mosquito could easily find soft spots to probe. Nothing came of that either. If a mosquito had fed, I would have known about it next morning, for the dark, distended abdomen would have been a sure indicator.

But there may have been one other reason why they would not eat. Most kinds of mosquito will not bite until after they have mated, and if these females were still virginal, we could not conclude that they truly disliked man, mice and chicks. Indeed they probably would have bitten them all, but were not yet ready for that signal on their computerized program. Mating was certainly a key to unlock other forms of behavior. But after that, the domesticator still cannot sit back at ease. Females do not necessarily lay their eggs in just any water container left about casually. Oh, no, some of the requirements for oviposition are just as strict as those for mating and feeding. Then a suitable larval food must be found, either a commercial one such as brewers' yeast or powdered dog chow pellets, or something that the entomologist concocts himself from materials found in the wild larval habitat. Difficulties present themselves in a continuing series. Whenever I read a report of the successful colonization of a new species of mosquito, I always lift my mental hat respectfully to the investigator who has been through all that hell.

Hugh Paterson's arrival from England occurred at exactly the time when I was ready to abandon *C. theileri.* I was licked and I knew it. But I could tell Hugh a few things. The insectary was primitive as regards regulation of heat and humidity. It was always too hot or too cold, and the only way we had of humidifying mosquito cages was to drape wet towels over them. By morning the towels would be dry, so that everything was fluctuating all the time. No wonder mosquitoes died! If we were going to make transmission experiments, we must have an entirely new plant, equipped with modern gadgets, in which mosquitoes could be reared abundantly and confidently generation after generation.

Hugh was delighted. "That is exactly what I have been saying all along," he said, "but no one would listen. They claim there isn't enough in the budget for anything as elaborate as that. With two of us now asking for the same things, maybe we can get somewhere."

We did. But since this was the kind of entomology Hugh liked, while I preferred Ndumu, I gratefully turned *C. theileri* over to this colleague and thenceforth watched him struggle with it when I happened not to be in the wilds. Trips to Germiston Lake also terminated for me. But I'll guarantee that if I walked into the clubhouse today, the manager would give me a martini "on the house." Perhaps two. And I think I'd deserve them.

12

SOME PEOPLE AND SNAKES
AT NDUMU

Though I sometimes felt that our planet had come to a halt
in its orbit and winter would never end, I was wrong. Blissfully I
found myself again identifying mosquitoes at Ndumu. The daytime examination of hundreds upon hundreds of specimens never
palled, though there *was* a certain monotony when most of them
were *Aedes circumluteolus*. The remainder consisted of a mixture,
including perhaps a dozen or more species, that nevertheless became quite familiar to me after I had seen them the first fifty or
seventy-five times. The real jolt would come when a stranger appeared. During the day I would not have time to track it down in
Edwards' book, for the routine of identification, killing and freezing known mosquitoes for future virus tests had to be consum-
mated first. But I would hoard the thought of that odd specimen
in my mind. This would be my reward—my after-dinner cordial.
When all other duties were out of the way, I would set up the
dissecting microscope, mount the mysterious specimen on a pin,
open Edwards at page 1, and savor the ultimate joys of entomology.

One evening Pottie and I were sitting in the screen-enclosed
laboratory room while I was thus occupied, when we heard a
commotion and babble outside. Before we could get up to investigate, Jack rushed in and began talking rapidly in Zulu.

"They've brought a woman who was bitten by a snake," said Pottie, "and they want us to take her in the van to the missionary lady for treatment."

The "missionary lady" was not a doctor, but she had a few beds for patients and ran a small dispensary in which she could treat minor injuries and infections. The mission was financed by some charitable religious organization in our Midwest and was supposed to tend primarily to the conversion of heathen souls. But the natives took small notice of that part of its function. Whenever they experienced a serious illness or accident, they turned Christian temporarily in order to receive the lady's ministrations, but, once again well, they reverted to paganism. The local witch doctor naturally told his fellow tribesmen that he was better than the missionary, but in general his practice embraced ailments for which physical remedies did not exist; and the lady was not one to indulge in ceremonies and incantations to restore or enhance virility or to exorcise devils (unless hymn singing can be classified thus). Consequently the mission and the witch doctor were not greatly in competition with each other.

The mission was very close to Dupe's store, thus lying six or seven miles from our NRC camp. It would have been a long trek for those who undertook to carry the bitten woman through the dark for such a distance. We had made it a rule not to treat people ourselves, for that would have put *us* in competition with everyone, especially the witch doctor, and might possibly lead to ill will among the local population by the time they had listened to enough antagonistic propaganda about us from practitioners jealous of their patronage. (However, just to make our blood-letting for virus studies legal, Ken, Bob and I had been issued restricted licenses to practice medicine in the Union of South Africa. These certificates stated that we were authorized to treat natives and missionaries only. *That* kept us out of competition with the legitimate medical profession, for neither natives nor missionaries had any money.)

Pottie and I went outside with flashlights and found a young

African woman lying on the ground. A group of natives, now muted, stood by in a kind of horror, as if they were watching the approach of death. Pottie asked them what had happened, and the babble broke out again.

"She went outside her hut to relieve herself in the bush," Pottie reported. "It was too dark to see anything, but she knew where she was going. She had just squatted down when she felt a bite behind her right knee. She crawled back to the hut, cried out that a snake had bitten her, and collapsed. Then they decided to bring her here."

I had never seen a case of snakebite and therefore took a close look at the woman. I quickly found a skin abrasion at the site she had specified. She was conscious and did not appear to be in pain, but she was definitely in shock. Her skin was clammy and her respirations were almost imperceptible. Such signs were not incompatible with poisoning by cobra venom, which attacks the nervous system predominantly.

Pottie told the natives to put the woman into the rear of the van and to come along with us to the mission so that they could carry her out when we arrived. The "missionary lady" had not yet retired for the night and was soon cleaning the wound with alcohol and swabbing it with iodine. Under better light I took a closer look at the lesion but could see no puncture marks such as would be made by fangs. The skin had been scratched—perhaps even pinched—but otherwise it looked almost intact and there was no undue swelling.

The "lady" finished her job by putting a Band-aid over the site. "She can sleep here, and we'll see how she is in the morning," said the good savior of souls. "Let one of her relations stay here with her. If she gets worse, we'll send her up to Ingwavuma in the morning."

"Up to Ingwavuma" meant the mission hospital there. The "lady" did not like to refer to it directly, for this was an institution far superior to hers, staffed with real doctors. She yielded patients to them reluctantly, for she could get in some of her best proselytizing when her charges were flat on their backs. Those that finally were taken "up to Ingwavuma," a distance of thirty miles or so over a rough mountain road, were sometimes the worse for their delayed transfer.

I suspected already that the woman had been bitten by a harmless snake, and that her physical state could be attributed to fright alone. Therefore I privately approved the Band-aid and the putting to bed on that occasion. However, that is the procedure this lady would undoubtedly have followed in a truly serious case, so I suggested Ingwavuma anyhow, just to instill conservative notions into the amateur clinician's head. But she remained firm and even promised to sit up with the woman, though I doubt she did. Next morning the woman walked the many miles back to her hut, perfectly well and still a heathen.

On another occasion I became really alarmed. A similar confusion outside roused Pottie and me after dinner and we found that one of our bird-netting attendants had indulged in a fight and been stabbed with an assegai. The two-edged blade of an assegai is long and narrow, so that it can penetrate deeply while leaving only a slight gash in the skin. This man had such a lesion, scarcely half an inch long, on the lower right side of his chest, overlying just where his liver must be. The edges of the wound gaped apart, bulging with subcutaneous fat. There was no telling how far the spear had gone. If the liver had been incised, the man might die of internal hemorrhage.

Pottie drove fast that time, though I cautioned him that bumps in the road might make the bleeding more severe.

"I'll put a Band-aid on it and keep him here for the night," said the lady. "We'll see how he is in the morning."

Nothing would dissuade her from that course. I was hardly able to sleep, but of course the Band-aid saved the day.

I had the feeling of living closer to snakes at Ndumu than in any other place in my experience. Not that an extraordinary number of people were bitten—though snakebite could not be considered a rarity either. One heard frequent reports of cattle dying from bites of the black mamba, a long, somewhat slender and definitely aggressive serpent related to cobras, although it has no "hood" to spread. Death was said to be extremely rapid following injection of a full load of venom—twenty minutes or at most half an hour, even if the victim was a large bull. Human beings succumbed even faster, if the local lore is to be trusted. Yet all the natives roamed through the bush barefooted and barelegged—in fact bare almost all over—as if danger were a part of fate and therefore to be dismissed from constant remembrance. They all killed every snake they saw, whether it was poisonous or not, and that may have kept the reptilian population under a slight check, though the chances are that reproductive pressures quickly replaced lost members.

I'll admit to having been more cautious than usual in my field work at Ndumu. Elsewhere I have shared the fatalistic view because the risk really is negligible or at least comfortably small. But here one saw snakes often, especially along roads and trails, so that the bush must have held many others.

Pottie, for all his great bulk, strength and fearlessness in other respects, was terrified of snakes. One of his chores was to gather firewood for Jack's stove, and to that purpose he would grab half a dozen mosquito catchers after the morning's work and drive them a few miles to a burned-over area where there were many fallen dead trees. Everyone would hack away at old trunks and branches, eventually filling the rear of the van so that the mosquito boys had to cling to its sides or ride on the fenders on the way back—which they loved to do. Wood gathering was always a party. In fact, Pottie made all work fun for them in some marvelous way that he alone knew.

Pottie arrived back from one such excursion trembling. "Doc," he gasped, "I almost wrecked the van. I was driving back and suddenly saw a tiger snake on the floor by my foot. I was so scared that I couldn't see. It's a wonder I didn't crash into the ditch."

"Where's the snake?" I asked.

"Oh, we stopped and the boys killed it," he answered.

"You should have brought it to me, you know," I reprimanded him.

"Oh, Jesus, Doc," he wailed, "I know I should have, but I was so goddam scared that I didn't think about it. It was right by my foot and I was driving and—"

"You've told me that already," I interrupted. "All right. Now, where do you suppose the snake came from?"

He felt that it must have been inside one of the log sections that did not require chopping. Otherwise it would surely have emerged before the wood was loaded. After that episode, Pottie tried to peer into every chip the boys brought to the van, but all the same he missed another so-called tiger snake some time later. *That* one came out in the house, and Pottie encountered it climbing up the screen door of the sleeping porch just as he was walking naked into the shower room.

"Oh, Je-*sus*, Doc," he yelled. "Come here quick!"

I never learned the true name of that snake. The only tiger snake I know about is the much larger one that lives in Australia, where it is the most deadly of its kind. Now I beheld a beautiful slender creature about four feet long with handsome black, red and yellow markings disposed in outlined diamond shapes, and with a triangular, arrow-shaped head that certainly bespoke a venomous potential. But all the snake seemed to want to do was to pass beyond the screen door—a perfectly reasonable thing for it to wish. While Pottie watched in anguish, I offered the snake the handle of a butterfly net to twine about and then carried it to the border of the NRC compound, where I flipped it into the grass.

Another wail! "Gosh, Doc, why did you do that? Now it will still be around here and—"

"Go ahead and take your shower," I said.

Pottie would have fainted if he had had one of my experiences. We kept a stack of old magazines on the floor of the latrine, and one morning I picked up a copy of *Time* to read for a while.

When I glanced up a little later, I found that I had uncovered a coiled snake that was now looking at me intently with its large dark eyes. It was of a steely-gray hue, without sheen, and generally slender, though about four or five feet long (which I considered pretty big at the time). The head, not wider than the body, encouraged me, but that was not a totally reliable sign, for many poisonous snakes are built on those lines also. The question: What to do? I was in a rather awkward position to act, so I decided to let the snake do something first. Perhaps it would interpret my stillness as the lack of threat, whereupon it could safely uncoil and make its exit through a crack under the door.

But it just lay there and continued studying me. It was not coiled in a position for striking—that is, with right and left kinks down the body. Rather it was disposed in a spiral, which is more the posture of sleep. The eyes were those of a nocturnal animal, and I imagine the snake had hoped to doze for the day under cover of the magazine.

Very slowly I rose and girded myself for what might come. As I began to take a step forward, the snake came to life. So sudden was its action that the crack under the door was overlooked. For a few moments the snake and I were fleeing from each other round and round the small enclosure. At one turning I flipped up the hook and we both burst out into the compound, the snake darting like a racer into a mealie field beyond the fence.

Pottie could afford to be wise and complacent about that one. "A rat snake, Doc," he said. "Perfectly harmless."

In contrast to Pottie's fear and the natives' usual resignation toward snakes, Mkawpis was reported to have killed one in a most reckless manner when there had been no need for him to risk his life. I have already stated that this older brother of the mild Dom-Dom was a wild man. Apparently he was a show-off as well. I did not witness this adventure, but Pottie was there and insists that it is true.

Some mosquito catchers, coming to work, spotted a large black mamba in a low tree overhanging the path. They were talking about it at the camp when Mkawpis arrived. He demanded at once to be taken to the tree, and naturally everyone went along to see what he was going to do. According to Pottie, Mkawpis scrambled up the trunk, shook the snake off the branch and jumped down after it so fast that he broke its back with his feet the instant after it had reached the path.

I maintain reservations about that mamba but not in the case of another one that Pottie struck with the field van near camp. He was driving as fast as the road would allow (no—faster than that, I guess, for he was always breaking springs), when we saw what looked like a fallen stick ahead of us. It was actually an exceptionally large black mamba basking in the early-morning sunlight. Too late it became aware of the onrushing vehicle. The "stick" sprang up and two great jaws, spread apart at one hundred eighty degrees, struck the right front fender. The bumper simultaneously hit the snake's body heavily. Pottie stopped when he could and then backed up to where the snake had been. We did not like to get out to look for it, lest it be in nearby grass clumps, injured but still able to attack.

Many people think that running over snakes with automobiles does the creatures no harm, as if they were made of solid rubber and contained no delicate viscera. The very opposite is true. The fact that the snake can still move is not a criterion of its having escaped injury. The chances are that almost all run-over snakes die eventually, but, being cold-blooded, take their time about succumbing. I have dissected several such "road-kills" and found crushing injuries of organs such as lung and liver, accompanied by severe internal bleeding, to be universal.

Pottie ran over a puff adder on another occasion. We could see it still lying in the road behind us. Again he backed up, and this time I went right up to the snake, for these thick, stubby adders aren't much when it comes to racing. It lay there as if nothing had happened, though I had felt the bumps when a front and rear wheel went over it.

"It must be dead, Doc," said Pottie from the safety of the van.

But it was not dead. After viewing me for a few moments, it moved slowly off the road into a thicket. In my imagination I could see the blood oozing from mutilated tissues and I foresaw the slow demise of the snake unless it was lucky enough to be dispatched sooner by a large predator such as an eagle.

At Site 26 along the Ingwavuma River there was a small pool where I sometimes dipped for mosquito larvae. It was a fine pool and provided me with exceptionally valuable material until I became afraid of the place. It happened to be the favorite resort of another hunter in the form of a green mamba. These absolutely beautiful bright-green snakes are said to be arboreal, but this one was often on the ground or climbing slowly on the rough stone face of the dam that caused the pool to exist. It was not a large animal—only about five feet long—and its attractive appearance and deliberate actions gave it an aspect of harmlessness. However, I was told that the green mamba is even more deadly than the black kind; hence my subsequent diffidence when it came to collecting mosquito larvae from that nook.

I have no statistics to prove the impression I gained that more native lives were lost to crocodiles than to snakes. You simply heard of more cases involving crocodiles. Most of the victims were women filling water jars on riverbanks. The Usutu did not boast many of those beasts, but they were present in sufficient numbers so that you could usually spot one or two from a lookout point called the Red Cliffs.

A man was thought to have been murdered in Swaziland and his body thrown into the Usutu to get rid of the evidence. Swaziland authorities notified authorities of the Union of South Africa, suggesting that the body should float downstream and pass Ndumu in about three days. Our local police and the game rangers alerted all natives to be on the watch, and exactly according to prediction the body was recovered.

"I'm going along to the Red Cliffs with the ranger to look at it, Doc," said Pottie.

I told him to go ahead, for I was busy sorting mosquitoes. An hour later Pottie returned, looking as green as a human being can.

"It was awful, Doc," he groaned. "I wish I hadn't seen it. The murderer had cut off his private parts, but that wasn't so bad. It was the head. Every crocodile along the way must have tried to swallow it, because you could see the rows of teeth marks all over. I had to vomit."

Pottie's recital so revivified the scene in his mind that he now had to go out and vomit again.

My mosquito catcher's title, "Qmba Ngwenya"—"bloated crocodile"—may have been more than an outlandish name. Perhaps it was believed to have some kind of protective charm. The same may be true when it comes to snakes, for an old African who took care of the NRC premises was named Mamba. Nothing funny about that one; it was not "crippled" "bloated" or any other kind of descriptive adjective—just plain Mamba.

Otherwise I do not remember hearing male names that referred to living things. Perhaps the girls were named for flowers or birds, but one did not have much contact with females here, and I believe even Pottie did not learn much about them in his long conversations with the natives—males, that is. If there had been anything interesting to report, he surely would have told me.

13

HONEY-GUIDES
AND GUINEA FOWL

We could not ignore any form of wildlife at Ndumu, in case it might harbor unsuspected arboviruses. One reason I had scolded Pottie for killing a snake was that I would have tried to get a blood specimen from it for the lab if its heart had still been pumping. Mammals, birds, reptiles and amphibians—we were covetous of them all and tested not only their blood but also suspensions of their assorted livers, spleens, kidneys and other viscera. Likewise we looked on biting arthropods very broadly, for anything that sucked blood might imbibe viruses at the same time. Thus my various non-mosquito activities, though they often looked simply like diversions, were never completely carefree. On every bird walk I kept alert for possible virus angles. But we could dismiss nothing—not even Mamba's house or Mr. Babcock's guinea fowl.

Mamba wore the most tattered and patched pair of knee-length khaki shorts I have ever seen. I believe this was the only garment he possessed. He lived on the compound in a small, tin-roofed, plaster-walled shack similar to Jack's except that it sported a small window, which, however, Mamba had boarded shut. After Jack had butchered his slain animal, Mamba bought several pieces and hung them inside on his wall. Perhaps the boarded window was to prevent breezes from diluting delicious aromas of decay.

Pottie took me to look inside the shack one day (after the door had been open for several hours). There was nothing in it but a few cooking utensils and a mass of old rags in one corner where Mamba slept. But Pottie did not bring me to see that.

"Look at all those long bloody streaks on the wall," he said. "Each one is from a full bedbug that Mamba squashed. I don't know how he can sleep."

Bedbugs they probably were, but an unpleasant thought that came to me was that they could be relapsing fever ticks. In either instance, Mamba's domicile was less than a stone's throw from our building and it seemed inevitable that some of his well-fed vermin must ultimately invade our premises.

Well, then, why not give the poor fellow's hut a good spraying with DDT? Why not spray our own quarters while we were at it, as a prophylactic measure? Alas that we were engaged in an entomological program! DDT even nearby would be carried on dust particles by the wind to my screened cubicle and put an end to the possibility of rearing mosquito larvae in jars and pans. Or else the mosquito boys would track DDT in on their feet when they delivered their boxes of filled tubes. DDT and entomology simply do not mix.

But Mamba seemed always smiling, cheerful and energetic. No bags under his eyes betokened night-long wakefulness. I don't know how old he was beyond that he must have been ancient. Oddly enough, considering the spoiled meat that he found so delicious, he also had a sweet tooth. At certain times of year (I think it was in the dry season) he went into the bush almost daily and returned with dripping combs of wild honey, invariably sharing it with us. The honey was dark amber and quite bitter, but it was nonetheless excellent—heavily sweet beyond question.

"Where does he find so many bees' nests?" I asked Pottie.

"Oh, *he* doesn't find them—he follows the honey-guides," was the answer.

Of course I had heard about honey-guides in a vague sort of way, and I remembered that ornithologists had originally dismissed as legend the reports of travelers who quoted natives as saying that certain birds would lead them to wild hives. But I recalled also that today the fact was supposed to have been verified. However, I was not prepared for Mamba to be able to go out and return so regularly with a harvest. I had thought that the phenomenon must at least be only occasional, if not rare, and that man might be sought less often as a follower than honey-eating wild animals such as ratels (or "honey-badgers") and even *they* might not wittingly follow honey-guides but find the hives independently.

Clearly I knew nothing about the subject if my thinking was as confused as that. I had already *seen* the birds at Ndumu, near Site 18. The common species here was the Greater Honey-guide, which has the suggestive Latin name *Indicator indicator*. Scientific titles do not always impress me, however, for taxonomists have often woven myths and fables into nomenclature. Our Chimney Swifts, *Chaetura pelagica*, for example, are certainly not seabirds. Their winter home in the Andes was unknown until recently. Prior to that, more than two hundred years ago, it was believed that the birds hibernated at the bottom of the ocean, and Linnaeus did not hesitate to immortalize so fantastic a notion in giving them their classical sobriquet (though Greek *pelagos* refers more to the ocean's open surface than to its depths).

Moreover, I had also heard the Greater Honey-guide give the call which, as I read in Roberts' book, invited me to a game of follow-the-leader. Again I was too skeptical to heed the note, though his description of it was apt enough: ". . . may be duplicated by rattling a half-empty matchbox *lengthwise*." Birds rattled at me on several occasions, and they must have concluded that I was one of the stuffier members of my species when the only response I gave was to rattle a half-empty matchbox lengthwise *at them*. They paid no attention to the sound.

Then, quite unexpectedly and to my great edification, I received a copy of Dr. Herbert Friedmann's recent and definitive volume called *The Honey-guides*. This came from Dr. Friedmann himself, with a request that I send him living specimens of those birds, if I could obtain any. He had studied several species in the bush himself, but as curator of birds at the U. S. National Museum in Washington, D.C., he could not spend all his time in Africa and had to depend on field volunteers for fresh materials.

I shall reveal Dr. Friedmann's motives in a moment, but first I must report that his book put the question of "guiding" on an incontrovertibly positive basis. Only a few species of this small family of birds have developed the guiding habit, but almost all of them are honeycomb eaters. It was believed originally that what they wanted was honey, but that turns out to be the astounding untruth. Dr. Friedmann and various collaborators established the fact that particular bacteria and yeasts (*Micrococcus cerolyticus* and *Candida albicans*) in the digestive tract of Lesser Honey-guides are able to break down part of the wax to render it fit for assimilation. But Lesser Honey-guides, despite their name, almost certainly *do not guide,* depending for their gratification on hives that have already been exposed by animals or human beings.

Guiding is apparently restricted to only two species, the Greater and the Scaly-throated Honey-guides. The scaly-throated one is the less common and the more irregularly distributed of the two, and I never identified it at Ndumu. But Greater Honey-guides were obviously abundant here, as attested by my seeing them sometimes and by Mamba's being able to gather honey with their assistance practically at will.

Dr. Friedmann now wanted to round out his study by testing the bacteria and yeasts of a *known* guiding honey-guide for their ability to split beeswax into digestible components. It should be possible for me to trap Greater Honey-guides for him, he said, by exposing pieces of honeycomb on exposed branches and arranging snares of some sort about them. The birds were known to respond to that sort of "feeder," though it seemed a far remove from the sunflower seeds with which I used to lure cardinals into my banding traps.

I had no spare time at Ndumu for conducting Dr. Friedmann's

project personally, but I asked Pottie to spread the word that I would pay for honey-guides brought to me alive and in good condition, stipulating that the birds must be caught *outside* the game reserve. The latter condition must underlie the reason why the final result was nil. Nobody wanted to go to that much trouble. I wondered also whether the natives might have been reluctant to trap honey-guides. After all, they could see usefulness to themselves in those creatures, and perhaps they liked honey better than money.

Mamba used to bring us other gifts from the wild at appropriate seasons. Among these was one that I think he prized even more highly than honey. Considering that it was only briefly available each year and was limited in quantity even then, I was greatly touched by his generosity. However, that was the only part of the gesture that moved me positively, for marula "wine" was simply dreadful. The natives made it by fermenting the plum-sized yellow fruits of marula trees. It ended up with a fairly high alcoholic content and an extremely dense turbidity, so that if you allowed a glass of it to settle, about 50 per cent of it turned out to be sediment. And it tasted like rotting fruit, which, of course, it was.

I have mentioned that natives were forbidden by law to drink. Therefore marula wine was never brewed openly. I used to see Mamba gathering fallen fruits under the marula tree in our compound, and I got as far as finding where he stored them in an unused shed at one corner of the property. But when he had accumulated a sufficient quantity after several days, both he and the marulas would disappear into the bush. From time to time he would come back to do some work, but his presence was intermittent. I suppose he had to keep attending to the mash.

Part of the recipe must have required constant tasting to judge the chemical progress, for when the wine was ready and Mamba returned with his containers, he was already hardly able to walk. Then for several days he would be on a splendid drunk and tell Pottie that he felt like a young man. I am not the only one who has observed that the natives at Ndumu seemed to be extremely responsive to almost neglible quantities of alcohol. That may not be entirely because they were unaccustomed to it. The chances are that their stomachs were never very full, and everyone knows that a cocktail has a much greater kick before breakfast than after dinner.

Mamba had several cronies, almost as old as he, who came to see him occasionally. Pottie often joined them in long conversations and would afterwards tell me what had been said. Apparently one way they had of reminiscing was not so much to talk about their youth as to repeat old tribal legends. They deplored the present young generation that seemed uninterested in memorizing those traditional recitations, and it is likely that many of the tales have now been lost.

Otherwise they spoke of life as they currently found it. One old man had an arthritic wife whom he had to carry everywhere on his back. He looked too decrepit himself for that sort of duty, and when I said as much to Pottie, he answered, "Just wait till next time."

"Next time" the old man came over to the front of our building and, at Pottie's word, put on a command performance. He suddenly went into a frantic and violent dance, ending it with a somersault in the air that landed him on his back.

Another old man confided in Pottie that he had lost his sexual vigor. Did we, by any chance, have any medicine that might help him? Pottie assured him that we did but that it was very strong, so that the tablet must be divided in half, only one half to be taken at a time. Later the man reported excellent results from the aspirin and almost wept when Pottie told him that he must wait a long time before it would be safe to refill the prescription.

An old man's club being in existence on the NRC grounds, Mr. Babcock fell easily into its membership, if you can call seventy-five "old." That gentleman, Mr. Orville G. Babcock, of Sonora, Texas, was Bob Kokernot's father-in-law and a retired government entomologist. He and his wife had come to visit their family in Johannesburg, but you could not expect an entomologist, old or young, to waste his first trip to Africa by remaining in the city. As soon as the next scheduled Ndumu "session" came up, Bob had no choice but to join the party and to bring Edith's irrepressible sire with him to "bug heaven."

For that is what Mr. Babcock found it to be. He was so stimulated to see an entire world of insects not one of which was familiar that he wanted to begin collecting everything at once. Naturally that led to a sort of catatonic state, out of which he soon emerged with a bit more sense of direction. His work had been concerned chiefly with insects and mites of veterinary importance, so now he cast about for ways of seeking prey of that sort. Here is where he teamed up with Mamba for a while. Pottie acted as interpreter, explaining to Mamba that Mr. Babcock desired to look for mites in the soil. I don't know how Pottie said "mites" in Zulu, or even whether or not *he* knew what mites were, but the message got across somehow and the two figures could thereafter be seen, on their knees, grubbing around the roots of trees, turning over old boards, and sifting humus and sand between their fingers.

"Very few mites here," said Mr. Babcock on the third day. "Soil's too dry, I guess. I saw some scavenger flies, though, and they ought to be interesting. Have you got any scrap pieces of screen wire lying around? I want to make a fly trap. And, oh, yes—I'll need some bits of raw meat to bait it with, the older the better."

So then we had a spate of carpentering or, if that is too inelegant a term, cabinet making, for Mr. Babcock was an expert in that line, too, having made all his insect trays and specimen cases at home in Sonora. Unfortunately we could give him neither proper tools nor adequate materials, and the trap he produced *was* inelegant. However, it caught flies, which was after all the basic intention, and Mr. Babcock was ecstatic.

Mamba had not been recruited for this project, so Mr. Babcock went it alone, hurrying out any number of times each day to see if anything new had entered, making careful mounts of such as did, and writing lengthy notes about each one.

So the idyll might have continued until the end of the ten-day session, had not Pottie struck a Crowned Guinea Fowl with the van. He brought the carcass to camp, observing sadly that there wasn't enough flesh on it to feed our entire party. If he had killed six, I still doubt that we would have seen anything on our plates, for Mr. Babcock at once seized the bird as if it were recently living gold.

"Mallophaga!" he exulted. "Feather lice! We have domestic guinea fowl in the United States, but most of their feather lice stayed behind when they were brought from Africa or else they have been lost since. Nobody knows how many kinds of lice guineas normally have in the wild. I'm going over that bird from one end to the other and collect every louse on it."

I could share his enthusiasm, for I had had a recent revealing "louse experience" myself. Lice are generally of two kinds, mallophaga (feather lice or biting lice), occurring principally on birds, and anoplura or sucking lice, confined to mammals. Mr. Babcock was actually interested in both kinds, for one of the high points in his life had been the discovery of a new species of sucking louse on the collared peccary in Texas.

These lice, of whatever category, are permanent ectoparasites, as well as being highly host-specific. Their transfer to new hosts is therefore an enormously critical point in the life history of each species. When animals or birds of the same sort come together for any reason, new infestations can be established without too much trouble, but if a host dies in solitude, its lice are bound to expire soon after the carcass has chilled. Should a mate come along immediately and brush the body, however, some of the lice might manage to clamber aboard in the nick of time, but that would be a rare happening. Probably many infestations are passed from parents to offspring, while the rest depend almost entirely for their perpetuation on social and sexual behavior, which, however, varies among the different vertebrate species to such an extent that each kind of louse must have its own novel problems to solve in the matter of achieving dissemination.

I recounted my "louse experience" to Mr. Babcock. One day when I was sorting a batch of mosquitoes, I had noticed a specimen of *Aedes circumluteolus* with two minute objects stuck to its legs. Looking more closely, I was amazed to see that they were sucking lice. ("Golly!" said Mr. Babcock.) Such creatures don't feed on mosquitoes. Then what in the world were they doing on this one?

I really did not have to ask that question, for I knew the answer at once. However, as I shall explain, this was nevertheless the next thing to a unique case. A scientific term known by very few scientists, including biologists, is "phoresy." This means the transport of one form of life by another for no reason other than to get the passenger from Point *A* to Point *B*. It is a well-recognized phenomenon among feather lice, which use hippoboscid flies as taxis. Most hippoboscids, also called louse-flies (though they are true flies and have nothing of lice in their genealogy), are capable of flying from bird to bird and thus serve as occasional supplementary means for mallophaga to invade new hosts. Three cases of mosquitoes carrying mallophaga are on record, but this is apparently a much rarer brand of car for lice to use.

So far as anoplura—sucking lice—were concerned, I could find only one published instance in which they had been caught practicing phoresy on a conveyance of any kind, though in that case the vehicle was a non-bloodsucking fly associated with cattle. Therefore my specimen of *A. circumluteolus* with its hangers-on was bound to make history. ("Well, I'll be!" said Mr. Babcock.)

I popped the burdened mosquito into a vial of alcohol and took it to Dr. Zumpt at my next opportunity. He turned it over to a colleague, Dr. O. Fiedler, and eventually I was informed that the lice were a species of *Linognathus* known from the red and the gray duikers, both of which small antelopes occur in the Ndumu Game Reserve, as I have said. Since sucking lice do not practice phoresy as a rule, I assumed that a duiker had died (or been illegally slaughtered) and my mosquito had alit on the carcass while it was in the cooling-off stage. Lice were already alert to the catastrophe and, instead of remaining next to the skin, had begun to abandon ship. These two succeeded in grasping a leg apiece and off they went for the ride of their lives.

Mr. Babcock really gave that guinea hen a going over. I don't know how many feathers a bird of that bulk may possess, but I do remember a paper by Dr. Alexander Wetmore in which he stated that he had counted nearly two thousand on a sparrow-sized Slate-colored Junco. Anyhow, Mr. Babcock wasn't going to risk over-looking anything, so he began taking that guinea apart *feather by feather*. Each feather had to be examined on both sides under a lens to be certain that no mallophaga escaped detection. Our general lab was soon a mass of feathers, and for months afterwards Pottie and I kept turning up those gray pennants with white polka dots in all sorts of odd places.

By the time the bird was naked (at least two days later), it was beyond cooking. Mr. Babcock was jubilant. "Our guinea fowl in the States have only two kinds of native African mallophaga on them," he said. "They sometimes pick up stray ones from chickens and even turkeys, but those don't count. Here, from this bird, I've got at least seven species, and I think possibly one or two others. My friends in the Department of Agriculture and at the U.S. National Museum are going to be mighty excited about this."

"Gee whiz!" I said.

14

IMMORTALITY

Eventually I found another specimen of *Aedes circumluteolus* that was being utilized as a louse carrier, this time for only one passenger, though it also was of the sucking kind. However, an entirely different sort of arthropod used mosquitoes commonly for transport. These were larval aquatic mites, so-called Hydrachnellae, and the chances are that they got on mosquitoes at the time when adults emerged from their aquatic pupae. The mites were not parasitic—that would have nullified the notion of phoresy, for this would have made them highjackers rather than mere stowaways. But since female mosquitoes are bound to return to the vicinity of water to lay their eggs, their exploitation by mites was a perfectly designed mechanism to assure those tiny arachnids opportunities to spread to all suitable environments.

It is amazing what will come out when you get down to fine points. After I had looked at enough mite-carrying mosquitoes, I became aware that just as the mosquitoes were of different species, so there was more than one kind of mite involved in this business, and, moreover, a particular sort of mite confined itself to particular kinds of mosquito. Thus *A. circumluteolus* invariably bore a very small, round, smooth, shiny, scarlet mite, or, I should say, a given mosquito sometimes carried as many as a dozen of them. The two species of *Taeniorhynchus*, on the other hand, were inhabited by a larger, slightly elongate, sparsely haired, less glossy, orange-red species. Larvae of *Aedes* and *Taeniorhynchus* have entirely different habits, those of *Aedes* being free-swimming and those of *Taeniorhynchus* attaching themselves to water plants. Obviously aquatic mites must be equally selective in the stations they occupy in pans and ditches. Yet both sorts have "discovered" the trick of utilizing mosquitoes for their dispersal.

Through my lens I could see that attaching itself to a mosquito must be rather a hasty act for a mite to perform. The emergence of a mosquito from a pupa is fairly rapid in itself, and it is not many minutes before the insect is able to fly. If this were not so, the mites could cluster in nice safe places like the sides of the thorax or the under surface of the abdomen. That, it seemed to me, is where they preferred to ride, for most of them did congregate in those sites. But the last ones aboard apparently grabbed the first surface available. I saw mites holding grimly to wings, legs (including the tips of the feet), antennae, the proboscis, and even the eyes.

Well, after you have examined 81,702 *A. circumluteolus,* as I did during two years' time, you *ought* to be getting down to fine points. And that is how I became immortal. On a January day in 1959 I was sorting mosquitoes as usual when I came across a specimen of *A. circumluteolus* that looked a bit odd. I wiped the tube carefully, saw to it that my glasses and my 5x eyepiece magnifier were clean, moved to the window for maximum illumination and looked again.

The mosquito was definitely "different," though at a quick glance it could have passed for *A. circumluteolus.* What drew my attention in the first place had been the uniformly yellow *unstriped* dorsum of the thorax, though that might still be a rarely permissible variation in a somewhat unorthodox member of the species. But at the limits of vision with my weak lens I thought I could see that the wing veins bore all sorts of yellow scales that far exceeded even marginal propriety.

I set the tube aside for further study later, because now my job was to get through with the rest of the day's catch so that I could put Qmba Ngwenya to work with the killing jars. But only a few specimens further on I came upon a second aberrant A. *circumluteolus*, if that is what they were. This was ridiculous. How could two look-alike freaks crop up on the same day? Despite the urgency of my work, I took enough time to compare the tubes side by side. There could be no doubt of the mosquitoes' resemblance to each other in every detail.

In all cases of questionable identification, or if I knew that a particular specimen was rare or perhaps needed in our collection in Johannesburg to complete an adequate series, I made pinned mounts of the mosquitoes, labeled them and stored them in special insect boxes where they would be safe from damage. That happened so infrequently that it made no lamentable dent in the numbers of specimens submitted to the lab for virus studies. Indeed it was part of the virus plan anyhow, for once the problems that such mosquitoes might present at first were solved, future captures of the same kind could be processed confidently in the lab's intricate isolating machinery.

My two mystery specimens continued to baffle me after I had them properly mounted and could view them under greater magnification with a binocular dissecting microscope. I tried to fit them into every couplet of Edwards' key to Ethiopian species of subgenus *Banksinella*, to which *Aedes circumluteolus* belongs, thinking that perhaps a more northern species had filtered down through the "tropical corridor" from equatorial regions and had just reached Ndumu. Some of those exotic species had unstriped thoraces, just like mine, but not one had wings with such gaudy veins. Though I hardly dared to believe it, I felt I must have discovered a new form of life.

Yet that would be a most unlikely possibility. This region had already been worked over thoroughly by three experts: Botha de Meillon, Jim Muspratt and Hugh Paterson. A novice following them was probably simply making a novice's typical sort of mistake. But I surely couldn't find the error. And I did bolster myself slightly by reflecting that probably all three of those entomologists put together had not looked at as many A. *circumluteolus* as had passed under my eyes.

This called for a conference with Botha in his Johannesburg laboratory. I told him directly what I thought, rather than challenging him blankly with the specimens, for I wanted him to take the defensive side at once. If I could crash that, everything would be fine.

He went through Edwards, just as I had done, but also brought out several trays of pinned Banksinellas from other parts of Africa. The collection at the South African Institute for Medical Research is excellent, most of it having been prepared by Botha himself. Among all this comparative material, perhaps he would find counterparts of the Ndumu wraiths.

He failed, and I waited expectantly to hear his corroborative words. "Very impressive," he said, "but not sufficiently so. Two specimens are very slim bases on which to describe a new species. If they really are something different, you'll be getting more of them. Just hold on to these and see if anything else happens."

"But I have heard of some people who have described new species on the basis of single specimens," I objected, trying not to sound plaintive. "Aren't two twice as good as that?"

"That's true enough," Botha agreed. "However, as you know, many described species eventually fall when it is discovered that they are only odd variants of something already known. That is especially common when the original series was too small. You can go ahead and describe these now if you wish—I certainly won't try to stop you. But my advice remains that you should be patient. If you get more, you'll be on safer ground."

"How many would you recommend?" I asked.

"Oh, I don't know exactly how many," he replied. "Just wait and see what happens, as I said before."

I could hardly endure the interval until our next scheduled expedition to Ndumu. At this session I must surely find additional specimens, for I reasoned that I had probably overlooked many of them before chancing to notice their slight but peculiar differences. It is common experience that we see best that which we are trained to see, and for the rest are sometimes as good as blind.

One might want to know why I was so keen to describe a new species. That was no mystery at all to me, so I can readily give the reasons. In my teens I had made my first contact with professional biologists at the Academy of Natural Sciences in Philadelphia. Although I did not realize it then, this was just before the end of a long era during which the chief function of natural history museums was to announce the discovery of new kinds of animals and plants to the world. Thus the people I met—and immediately worshiped—were practically all of the "old fossil" school, some of them with hundreds of discoveries to their credit. When these scientists had been relatively young men, each had chosen a certain fairly restricted field in which to specialize, so that he could entertain reasonable hopes of becoming an authority in it on a hemispherical or even global basis. There were Henry Fowler with his fishes, Ezra Cresson with his wasps, Francis Pennell with his figworts, Henry Pilsbry with his mollusks and Morgan Hebard with his grasshoppers, for example. Who, in his young right mind, would not want to be like those masters?

Several concurrently developing factors combined to push that kind of natural history into the background. Mind you, it is still going on, as it surely must, but it is now far from the dominating urge in the progress of a good natural history museum. One of the most obvious causes of the decline in emphasis on taxonomy was that several centuries of concentrated work had led to the discovery of almost all conspicuous animals and plants. Probably a million or more forms of life remain to be named, but they become, with each passing decade, progressively smaller, less important to mankind, more difficult to locate and exasperating technically to study. The only reason that so much is known about the microscopic characteristics of the sexual appendages of male mosquitoes is because as a race we are vitally interested in that group of gnats.

But the world was now moving along faster than it ever had

before and biology suddenly flowered into a living science. Ecology and ethology were born. People wanted to know how organisms managed to get along in their environments and how they behaved. When such persons with enough curiosity went to natural history museums to find out, all they saw were glass-fronted cases containing musty, stuffed specimens, and there was nobody visible to tell them anything. The authorities were in that sacred precinct called "behind the scenes," and if one did manage to make his way past sentries to any of those cubicles, the incumbent savant was likely to know very little, if anything, about the habits and associations of his special subjects. "Only thing I can tell you about this specimen—" he might say— "it arrived from Sumatra in a bottle."

Museums began to go through a thorough metamorphosis, both externally and internally. The old stuffed specimens in glass cases were replaced by the "habitat group," which displayed an instant in the lives of animals and plants at a definite time and place. As "old fossils" died, they were likely to be replaced by young Ph. D.s with training in animate aspects of biology. No longer did they remain behind the scenes all the time. Instead, one could often find them lecturing to lay audiences or in conference with representatives from industrial and pharmaceutical firms or government agencies. These changes had resulted further from a redistribution of money. Patrons rich enough to support private institutions fully became scarce. Museums now had to improve their public images in order to command admission fees, while giving a more practical direction to their research (though it toppled many an ivory tower) drew subsidies from big business.

I recognized all this as good, wise, sound—and inevitable. But I regretted it and still do in a purely personal and secret way. To me it seemed as if the old museums had been destroyed. Surely the pure academic atmosphere would never obtain in them again, for applied science must be served too. While it does no good to weep for the dead, one nevertheless indulges in it. And so, with all my good sense telling me that it was utter nonsense, I desperately wanted to discover and describe a new species of some kind on my very own, in the same spirit that my early heroes had done so.

A good dash of vanity was involved as well. To illustrate: If I were to refer to *C. theileri* formally and in full, I should have to write the name as "*Culex (Culex) theileri* Theobald." Where does Mr. Theobald come into it? He, of course, is the entomologist who published the first recognizable description of that species, and his name will be linked with *C. theileri* forever. It is rather ironic that a man can do much greater things in our society and be quickly forgotten although his work remains permanently with us. Perhaps we ought to adopt rules of language or convention so that one always said: "I attacked the can of beans with a can-opener Smith," or "I fastened the baby's diaper with a safety-pin Jones." Anyhow, you can see where my two little mosquitoes might lead, provided I could muster a few others like them.

Theobald, by the way, was very famous indeed. I made a tabulation of all the species of mosquito that ABVRU's various workers had encountered in Natal province and found that among a total of eighty-nine Theobald's name followed forty-two. His successor at the British Museum (Natural History) was Mr. Edwards, whose book was my mainstay, and that gentleman did very well, too, with twenty-seven species bearing his stamp. The remaining twenty were credited to almost as many people, although Botha was associated with three of them and Jim Muspratt had described two. Even good, venerable Linnaeus was represented by *Culex pipiens*, the common household gnat. Obviously I was hoping to invite myself into a sparkling company.

Mosquito sorting took much longer henceforth. I lingered over each specimen of *A. circumluteolus*, and I drove the boys mad with my insistence that the tubes be absolutely gleaming. On the first return session at Ndumu, I rapidly fell into despondency, for my hypothesis of having previously "overlooked" the supposed new species fell flat. Each *A. circumluteolus* was exactly as it should have been, with boldly black-and-yellow-striped thorax and conservatively marked wing veins. During that session, and in others to follow, A. *circumluteolus* passed through my hands one by one, building up to their thousands upon thousands as I approached my eventual total of 81,702, but, to use Botha's expression, "nothing happened."

I had almost resigned myself to his inference that my two pinned novelties might not really be "something different," but only strange variations from normal A. *circumluteolus* stock. However, if that were true, I should be finding intermediate examples. If dogs and cats were all of one species, the two forms that we know by those names representing only the extremes of possible variation within the genetic pool, we would constantly encounter all degrees of doglike cats and catlike dogs, at the median point finding a dog-cat that we could not ascribe to either side. But A. *circumluteolus* was either pure cat-cat or pristine dog-dog, whichever you prefer: no genetic spillover showed up anywhere along the line.

Then in November, ten long months after the original discovery, Dom-Dom brought in a box of tubes that contained two additional specimens of the marvelous creature. One was moderately worn, but even that was advantageous in a way, for despite such an esthetic drawback it was still readily identifiable as an obvious sister belonging to what had now become a quartet that stood apart from A. *circumluteolus* with all the dignity of self-determination.

"You can't miss," said Botha after he had studied the new pair and compared them with the originals. "Of course they are all females. It would preferable to wait until you get some males and, better yet, larvae as well. But that might not happen in the limited time you will still be here, and these four, conforming to each other as perfectly as they do, should be enough to convince anybody. At least *I* am convinced. You might as well go ahead and publish."

Did I "go ahead"! I wrote the paper that day and it appeared not long thereafter in *The Journal of the Entomological Society of*

Southern Africa. Of course I had to find a name for the new species, but that was easy. Golden-scaled wing veins were its outstanding characteristic, so it must be *aurovenatus.* As taxonomists do so often, they had recently discarded the subgeneric designation *Banksinella* in favor of *Neomelaniconion* which was belatedly found to have priority in published entomological literature. Thus I came to occupy an immortal niche with the following inscription: *Aedes* (*Neomelaniconion*) *aurovenatus* Worth. Greetings, Mr. Theobald! Greetings, my deceased mentors at the Philadelphia Academy!

Of course it is not quite as cut-and-dried as that. Publication serves only to annouce that a scientist believes he has discovered a new species. If no one accepts the discovery, and particularly if the most renowned experts actively repudiate it in print, the bright little bubble soon bursts and no trace remains except for a couple of wasted pages of type in an old journal.

But how were the experts to judge? They wouldn't all converge on Johannesburg just to peer at four tiny mosquitoes that I had been arrogant enough to endow with a golden name. Botha discussed the problem with me from several standpoints. One must distribute these specimens, not only so that they would be accessible for study by other scholars, but also as a matter of prudence. For example, during World War II several important museums in Middle Europe had been destroyed, and irreplaceable "type" materials were lost. "Types" are the specimens on which a scientist bases his original description of a species. He actually selects a single individual and calls it the type, while if he has a series of individuals, the remaining ones are called "paratypes." All of these then become permanent standards, like that international yardstick, the platinum meter, in Paris, and every subsequent comparison must be made in reference to them.

"I suggest that you leave the type and one paratype here in the collection of the South African Institute for Medical Research," said Botha. "But you better deposit the others somewhere else. This place could burn down, you know. Why don't you send one to Peter Mattingly at the British Museum (Natural History) and the other to Alan Stone at your U. S. National Museum?"

"Do you think they would really want them?" I asked.

"Of course they would," said Botha. "They'll be very glad indeed."

Mr. Mattingly wrote me a very kind letter of acknowledgement, as if he truly welcomed the arrival of that small package. "Generosity" was one of the words he used. As for Alan Stone, he was an old friend, and I had decided to wait until my return to the States, when I would give the paratype to him in person. *That* small package traveled in my suitcase in a long roundabout way as I made the journey home from South Africa via India, Malaya, Japan, Alaska and California. Not many mosquitoes have flown that far.

As soon as I could, I hopped a train from Philadelphia to Washington and made my way to Alan's office in a wing of the National Museum. Alas! he was on vacation. I carefully undid the wrappings of the box and peeped inside to be sure that the mosquito was still in one piece after all its peregrinations. Finding that everything was in order, I entrusted the rewrapped parcel to an assistant who promised to bring it to Dr. Stone's attention as soon as he returned.

Several weeks later I received his confirmation of receipt, but my dreams of immortality were suddenly shocked into that aftermath of most dreams: disillusionment. Alan had studied the specimen but could not see how it differed from *Aedes* (*Neomelaniconion*) *luteolateralis* (another species in Edwards' "*Banksinella*" key, though I had checked that one and decided against it). What, asked Alan, had led me to think that the specimen was unique? Could I send him a reprint of my description?

My reprints were all boxed up and on shipboard, being transferred from South Africa to my next station in Trinidad. As I sat in Swarthmore and read Alan's letter, all I could rely upon was memory. What would be the most telling point I could mention?

Finally I settled on vein 3, a short vein near the wing tip. This one was dark-scaled in all other Neomelaniconions but stood out particularly brilliantly in *A. aurovenatus*, with a thick double row of golden scales along its entire extent. I posted the letter and then waited. This would be the ultimate test, for if Alan remained unpersuaded, the rest of the world would quickly accept his denunciation.

But all was well. "I overlooked that vein," he wrote.

15

EXPEDITION TO LUMBO

On several other occasions I thought I might have discovered new species, for I would know immediately that the mosquitoes were not in the collection that I had studied so assiduously at the lab. I was right on the latter score, but a perusal of Edwards' book would lead me to acquaintance with forms already known or, in more difficult cases, Botha and Jim ran them down to established names. The sport was none the less stimulating, however, for now I had arrived at the state of the bird watcher who has learned his local fauna by heart and henceforth lives only for rarities. All told I ran across five species that had never been recorded in Natal province, and three of these were "new" to the Union of South Africa. The most remarkable one was named *Ficalbia* (*Ficalbia*) *circumtestacea* (Theobald—of course). Edwards gave its known range as "Sudan and Sierra Leone." That did not mean that my specimen had flown across half a continent—it had probably been bred right at Ndumu. Rather, it might be rare, or it might have habits that led to its being only seldom captured by entomologists.

On home ground at our field station I could play this game with great satisfaction and pleasure. But now I was in for a new kind of experience, and it was by no means a game either. Part of

ABVRU's program included serological surveys of the inhabitants in various parts of southern Africa, not only in the Union but also in neighboring territories. One such province-wide survey had been made by Bob Kokernot and Botha de Meillon in Moçambique (Portuguese East Africa) in 1957, the year before I joined the team. Laboratory studies on the serum samples disclosed some mysterious, unidentifiable antibodies in residents of the town of Lumbo, on the north coast. That is the sort of lead ABVRU was looking for. Having found it, we must now organize an expedition to make more detailed studies of the Lumbo region and its associated people and wildlife.

However, you can't simply walk into a foreign country and begin bleeding people and trapping animals just because you are interested in their viruses. But if you go about your plans diplomatically, doors are almost invariably opened to their full extent. Whether the results will be immediate (which is unlikely) or remote (even very remote), the outcome of such work will be for the public good. And if the country is poor, as is usually the case, it could not afford to conduct this kind of expensive study out of its own budget. The kinds of investigations ABVRU often made could be classified as luxuries, in a way, for they were directed toward situations in which threats or emergencies were not known to exist.

The survey of 1957 in Moçambique had opened such doors and nobody had dreamed of closing them since. Therefore it was now a simple matter to approach the same authorities with proposals for a more thoroughgoing study at Lumbo. We would be glad to accept whatever facilities they could supply locally, while ABVRU would furnish all other requisites. I had nothing to do with the arrangements, thank goodness, but I was none the less appreciative of all the groundwork laid by everyone else. Until close to the time for our departure, I simply kept at my routine mosquito studies in Ndumu.

On the day of our departure from Ndumu, an unusually large crowd of natives gathered to see us off, for we were taking with us one of their young men to help us in the field at Lumbo. This individual, Jay Mbuzi, was a favorite of Bob's for the irrelevant reason that his blood had once yielded an isolate of West Nile virus. That did not make him a better mosquito catcher or general worker; it did not give him greater initiative; nor did it mitigate the impression of shiftiness or slyness that he always made me feel. Yet Bob wanted him, so that was that.

As Pottie drove the van out of the NRC compound, Jay leaned out, shook his fist and shouted angrily, in Zulu, at one of the men standing by. Everybody, including Pottie, burst into laughter.

"What was that about?" I asked.

"These people don't care what they say or who hears it," answered Pottie. "That man has been after Jay's wife, so Jay just warned him, 'If you fuck my wife while I'm gone, I'll beat you up when I get back.'"

Nothing shifty or sly *there*, to be sure.

Since our expedition was to last for an entire month, preparations were correspondingly elaborate. Thus it was four weeks before we at last took to the air at Joburg and then several additional days were devoted to official duties in the capital, Lourenço Marques. All this time I was, as usual, itching to get to work, but one must first call on the provincial governor and thank him for his country's hospitality, and naturally it was stimulating to meet personnel of the Instituto de Investigação Medica, especially Dr. Alberto Soeiro, the director; Dr. Agosto Tito de Morais, the dashing young sparkplug of that organization; and Mr. Mario Pereira, resident entomologist.

Tito had already been to Lumbo to set things up for us. Jay Mbuzi had joined a contingent of Portuguese assistants assigned to the project and proceeded with them by boat from Lourenço Marques to the island of Moçambique on the northern coast, just off Lumbo, along with all our camping and scientific supplies as well as the field vehicles. These were then ferried to the mainland and disposed in the compound of the local *chefe de pôsto,* who in provincial villages is a sort of army commander, police chief and magistrate all rolled into one.

David Davis had sent along two of his native mammal skinners from Joburg as well, so that when we four arrived in Lumbo— Bob, Tito (who returned to supervise our settling in), David and I—all was in readiness. One tent had been set up as a kitchen, another for blood work, a third for David's dissections and museum preparations, and my own for counting bristles on gnats. We found Mr. Jacinto de Sousa running the show according to Tito's prior instructions. Jacinto had been "borrowed" from the trypanosomiasis (sleeping sickness) mission for the duration of our expedition. Finally, there was Mr. B. Rodrigues, in charge of the vehicles, as well as animal attendant and general handyman.

Not only did we drop into a perfectly organized physical plant; the wildlife was waiting for us too. Jacinto's first words were, "The mosquitoes are terrible here. There are a few all day, but about four o'clock in the afternoon they invade camp in such large numbers that we have to use repellant."

It was *then* almost four o'clock in the afternoon, so I hastily unpacked the boxes of mosquito tubes, set up my dissecting microscope and laid open Edwards at the ready. What would they be? *Aedes circumluteolus?* One of the common species of *Taeniorhynchus?* My old Ndumu friend, *Culex univittatus?*

Jacinto had already hired some local boys to catch mosquitoes, and although they had not expected to begin work until tomorrow, they were hanging about watching our strange doings, and might as well be put to use at once. Within half an hour I had two hundred seventy-six tubes on the table before me, each containing a buzzing insect.

Here is where my "new experience" began. I can still feel the shock that overcame me when I discovered that the umbilical cord leading to our mosquito collection at the lab had been cut. At Ndumu I fancied myself as an accomplished entomologist, but I was still dependent there on studies that had been completed by my predecessors. Now, in Lumbo, I was suddenly faced with the need to make decisions of my own, and I found that I couldn't! Only seventeen of those mosquitoes looked at all familiar. The rest, clearly all belonging to the same species, had me absolutely stumped. These were the common camp-biters. They were fairly large, with prominent cream-colored patches on the abdominal segments and a sort of dark wine-color suffusing the rest of their anatomies. I went through Edwards from cover to cover without being able to place them. Yet a form as abundant as this must surely be included in the volume somewhere—it could not have been overlooked by previous collectors.

The closest I could come to an identification was *Culex decens*, a rather nondescript species, but I was far from satisfied with that diagnosis. Daylight was fading. Bob said it was time to go to the hotel, so I disconsolately closed shop, thinking that from the hotel I might prudently continue back to Johannesburg next morning. Surely I was a failure as far as this expedition was concerned.

Rodrigues and some of our African helpers actually *camped* at the camp, while we spoiled scientists luxuriated in hot showers, waiter service on cloth-covered tables in the dining room, and civilized beds. The hotel existed as a facility of the railway terminal and was run by Mr. João Branco, an extremely hospitable Portuguese. Bob, David and Tito shared one large bedroom while Jacinto and I doubled up in a smaller one. Jacinto woke up more slowly in the mornings than anyone I have ever seen. He would get on his feet and pick up either a hairbrush or a razor blade sharpener—whichever came to hand first. Then he would brush and brush, or crank and crank, until in about fifteen minutes he became aware that that was what he was doing.

These were fine diversions, but they all seemed to take too much time. I wanted clock hands to spin in a blur until I could get back at those mosquitoes. For now there must be an end to nonsense. The species *must* be in the book; therefore *I* must be making some mistake. Well, then, let's begin on page one again and try each and every description to find one to fit this mosquito, even when it seemed preposterous (for example, taking speckled-winged forms seriously, when these specimens obviously had solidly dark wings).

That procedure, though tedious and blind, was bound to pay

off. When, on a most unexpected page, the puzzle solved itself, I discovered simultaneously what error I had been repeating. A general rule of culicidology is that females of *Aedes* species and those of a few related genera have pointed abdominal tips, while other mosquitoes, including *Culex*, possess rounded ones. However, here was the inevitable exception. My specimens ended in indubitably curved, rather than sharp, extremities, but they were none the less certainly *Aedes* (*Skusea*) *pembaensis* (Theobald). Perhaps their excuse for departing from custom lay in the fact that they were rather out of style in Africa anyhow. *Aedes pembaensis*, according to Edwards, is the only representative of the subgenus *Skusea* on this continent, all its relatives being Asiatic. Moreover *A. pembaensis* is found on the eastern coast of Africa, as if it were still only gaining a foothold here. It breeds chiefly in saline habitats (I now quote from other authorities), especially in crab holes. On Pemba Island, near Zanzibar, it is one of the commonest household mosquitoes.

Well. That was a good morning's work, and now my spirits were restored to such an extent that I heralded the mosquito boys' arrival with a surge of my old enthusiasm. This was a fantastic situation: of 690 mosquitoes they had caught, fully 654 were *A. pembaensis!*

Now I could afford to raise my head and look about camp to see what my colleagues were doing. Naturally it took hours to sort all those mosquitoes, but after the task was finished, I could drift over to David's tent or peer over Bob's shoulder. Among other actions that Tito had completed prior to our arrival was the broadcasting of an appeal for live creatures of almost any kind, particularly mammals, birds and reptiles. Through the *chefe de poste*, local residents were advised that we would pay cash according to the size, desirability and robustness of the specimens. (Moribund animals would be of little use to us, for they might not yield adequate blood specimens, and besides, we might in some cases want to keep the animals alive in order to take them back with us for further studies in the laboratory.)

Such requests for public cooperation rarely meet with much of a response. Remember, for example, the honey-guide bounties that I never paid in Dr. Friedmann's behalf. Foreseeing that kind of apathy, we had brought along several large packing cases containing a variety of mammal traps and bird nets as well as guns, ammunition and other collecting paraphernalia.

Most of that equipment was never unpacked. On the very first morning we beheld the unbelievable spectacle of a whole lineup of indigenes, each carrying a homemade cage containing a specimen of some kind. Indeed, some of the cages, containing vervet monkeys or yellow baboons, required two porters for their transport. One day they even brought in a jackal. Among smaller forms David identified two species of mongoose, a polecat, a bush baby (or Galago—a primitive primate), a palm squirrel, three kinds of rat and a shrew. Wild birds came in many sizes and colors and in more than a score of species. Small boys seemed most adept in catching these, while the men went after more challenging game. I forget who brought in the sluggish chameleons, but it ought to have been the old ladies.

Bob's training as a veterinarian as well as doctor of human medicine became a notable asset at this time. He was a whiz at knowing where to direct needles to copious sources of blood. According to the size and species of each animal, he might seek out the heart or else one of the peripheral vessels, especially, in larger creatures, the femoral artery. If a specimen was still drawing breath, Bob would draw blood.

It is virtually impossible to draw blood from a dead animal after the circulation has stopped, so we wanted to receive all specimens alive. But later we sacrificed them, even if they survived cardiac puncture, in order to obtain fragments of heart, lung, liver and so on for freezing and ultimate mouse inoculation, because rarely viruses can be isolated from solid organs even when the blood yields none. And as for slaughtering all those ill-treated creatures, we at least wasted almost nothing by the time David and his helpers finished preparing their skins and skeletons as scientific mementos. In fact Smithburn's Law extended to this arena also. Without an identified stuffed rat, mongoose or what-not to go with a future "positive" blood specimen, what could we say about the laboratory report other than that it referred to something that had happened in Lumbo? Moreover, in this kind of study it is informative to know the source of negative blood specimens in the environment as well. All types of findings put together may eventually reveal a pattern that means something to an epidemiologist.

Our collaborators stood about while Bob processed one animal after another, apparently fascinated by the esoteric procedure. But they had other motives, too. What about the promised money? This could not be paid until Bob had verified that each animal had been delivered in fit condition. At last, by morning's end, the treasury was opened and each deserving hunter received his reward.

The populace became so avaricious that some individuals took to stealing their neighbors' pets and bringing them in for sale. It would probably have been of scientific value to test blood from a series of local puppy dogs and pussycats, but the first weeping little girl who arrived with her family in search of a missing scrawny kitten demonstrated that it would be poor policy to exploit ambitions in that direction. Bob issued an edict that only wild animals would be accepted.

However, it is likely that we were still hoodwinked occasionally. Rodrigues informed us that one of the surviving banded mongooses in his care seemed tame, while the others were uniformly timorous or hostile. I went with him to look at the animal. When he put his finger through the wire mesh, this mongoose rolled over on its back so that he could scratch its underparts. No one had come to claim a missing mongoose, but I could not believe that this one was unfamiliar with human beings.

The upshot was that Rodrigues soon had the animal out of its cage and it became the camp mascot. He named it Lumbo. A banded mongoose is more handsomely marked than most other kinds, having about eight or nine narrow black stripes crossing its brown back transversely in the same pattern as the rings of an armadillo. Lumbo was not quite fully grown, but his variegated coat was already striking.

A pet mongoose I had had in India impressed me as being extraordinarily affectionate. Now that Lumbo turned out to be equally demonstrative, I concluded that this was not extraordinary but simply the way all tame mongooses respond. If Lumbo was not eating or playing, he wanted to sleep in contact with one of us—anyone would do, for he did not seem to discriminate among our odors or personalities. The main requirement was that the person must remain still. Thus, when I was sorting mosquitoes, he would crawl up inside one of my trouser legs until he reached the crook behind my knee where he could dispose himself on a cloth shelf and doze until I had to move. Then, when at last I stood up, he became greatly annoyed, though he never bit, sliding down my leg reluctantly and looking frowzy as he emerged, blinking, into daylight.

His play was of his own devising and amused us endlessly. One day he found a discarded rubber stopper from a broken glass vial. Henceforth this was his special toy. He would carry it until he found a vertical surface such as a wooden crate resting on the ground. Turning his back to it, he would project the stopper between his hind legs, exactly as a football center snaps the ball to the rear. Lumbo did this with such force that the stopper would bounce off the backboard for several yards, sometimes landing on our work tables. He could never see where it had gone, but he searched until he retrieved it (or we gave it to him), when he would scurry back to the crate and fire again.

At least the rest of us thought the gamboling was play until David made a most bizarre statement. "Lumbo thinks it's an egg," he said. Not that anyone had ever heard of rubber eggs. But David remembered a discussion he had heard in his student days at Oxford concerning the method mongooses universally use for breaking the eggs of wild birds. Though a mongoose can be a ferocious little animal, its teeth are relatively small and short and the gape of its mouth is not wide. Therefore it cannot get a grip on smoothly curved shells and must find other means to get at the contents.

"They said that mongooses break eggs just as Lumbo is trying to do with that stopper," said David. "The point being considered was whether a mongoose has to learn the trick or whether it is instinctive. If Lumbo was hand-reared, the chances are that he has never been given an egg. Yet even this stopper, which is only remotely egg-shaped, has elicited the 'breaking' response."

"Well," said Bob, "let's give him a real egg and see what happens."

There wasn't an egg in camp, but we sent off several mosquito catchers to buy some from local villagers. Bob set up his movie camera. Rodrigues placed an egg in front of Lumbo and *zip!* it flew between his hind legs and smashed against the crate before Bob had time to press the button. The rest of the eggs were likewise so speedily broken that nothing was ever recorded on film.

One day Lumbo ran away. Rodrigues was particularly distressed, for he had planned to take the animal back to Lourenço Marques for his children. He wandered about the compound, calling "Lumbo, Lumbo!" but no little mongoose came running. At last he concluded that one of the mosquito boys must have taken it, or it might have been killed by a dog.

But, after all, Lumbo had only run away. On the third morning Rodrigues found a group of four banded mongooses playing in a brushy area not far from camp. "Lumbo?" he called tentatively. At once three mongooses scrambled out of sight while the fourth scampered to his master's feet. This was a strange case of an animal's exercising prerogatives of freedom and captivity simultaneously, as Elsa, the famous lioness, was to do later. Lumbo finally made the journey to Lourenço Marques, for Rodrigues now insisted on keeping him caged, much to everyone's dissatisfaction— and that included poor Lumbo above all others.

Though we spent three weeks at Lumbo during which I did

little else than sleep, eat and sort mosquitoes, I found time now
and then to look at the surrounding countryside, such as it was.
Lumbo itself is situated on a rather narrow, slightly elevated strip
of land that faces a broad, shallow bay to the east. At the horizon
lies the island of Moçambique, which used to be the fortified capital of the province, plugging the entrance of the bay like a loose-
fitting stopper in a bottle. Behind Lumbo, to the west, is a man-
grove-bordered tidal lagoon over much of which I was able to
walk (or slosh) between its inundations. Here I found plenty of
crab holes, if that is what A. *pembaensis* needed. Crabs came in
many forms, from hermit crabs that did not dig holes at all but
lived in old snail shells, and armies of fiddler crabs that dug small
holes in the mud, to a larger species several inches across the back
and bearing red claws, that made deep excavations among the
mangrove roots, almost wide enough to admit one's hand and arm
if one were so inclined. Here, it seemed to me, were nurseries in
abundance for A. *pembaensis* larvae, though it remained mysterious why a species of mosquito should choose such out-of-the-
way places for hatching its eggs. Possibly the larvae enjoyed concealment from predators such as small fish in those retreats.

The exposed flats were not the most enjoyable places to stroll,
for the inhabitants of Lumbo used them as a community latrine.
If they had had much refuse to dispose of, the area would probably have become a dump as well, but fortunately the people
were too poor to live out of tin cans, and such bottles as they
possessed were kept for further use instead of being discarded.
Therefore the tide, varying in the extent of its daily incursions,
sooner or later rose high enough to flush out most of the wastes,
leaving behind only the soon-skeletonized carcasses of dogs or
hogs.

Inland from the lagoon the terrain was occupied by bucolic Africans who cultivated small plots of land, chiefly for their own use.
Rice, coconuts, cassava (manioc), plantains, cashew nuts, peanuts
and mangoes in season appeared to sustain them, apart from additional contributions by self-foraging chickens and goats. They

were undoubtedly only marginally alive, though to them it probably seemed a normal state.

In order to sample the mosquito fauna as thoroughly as possible, and also to collect insects in areas close to human habitations, where virus transmission might be a threat, we established fixed collecting areas which our catchers visited on a daily rotational basis. The first was in the environs of our camp; the second lay at the head of the lagoon, where women gathered to fill their cooking pots at a couple of freshwater pumps; the third was one and a half miles inland in the agricultural area; and a final one, four and a half miles inland, was centered in an old coconut plantation.

These sites quickly afforded us further insight into the natural history of *A. pembaensis*. I have already mentioned the prevalence of that species at camp and linked its abundance there to proximity of the lagoon and its crab holes. If a mosquito is born in that place and succeeds in finding a warm-blooded host nearby, there is small reason for it to leave the area unless it be helplessly carried away on a strong wind. It remains possible, however, that mosquitoes can be produced in such huge numbers that some of them fail to locate an immediate source of blood. Or perhaps, if flight is random in all directions, some of them simply are unlucky hunters at first and disperse themselves aimlessly until chance brings them to a suitable host.

The area of the water pumps more or less duplicated what we had observed at the camp. Here our catch of *A. pembaensis*, over the entire three weeks, amounted to 80 percent of all mosquitoes taken. However, only a mile and a half distant, in a patch of bushes and small trees near a cluster of palm-thatched huts, the species accounted for only 57 percent of the total, while in the distant coconut grove *A. pembaensis* had dwindled to a mere 21 percent. This decline in abundance may have been more apparent than real, for as we progressed inland we found correspondingly greater populations of more terrestrial mosquitoes, and *A. pembaensis* was simply diluted among its dry-land-frequenting relatives. However, one day I took Jay to a point twenty miles up the road to test that very hypothesis, and he found none of our camp pests whatsoever. *Aedes pembaensis* has been known to breed rarely in non-saline situations, but the evidence (which we were now able to strengthen) is that it would not last long if tidal estuaries disappeared.

Jay was not much help to us, just as I had feared. On the other hand, I cannot blame him too severely. It was simply that he took advantage of minor disabilities to shirk his duties longer than necessary. On the day we looked for *A. pembaensis* in the interior he was stung in the eyelid by a wasp. The eye swelled completely shut, and he now refused to use his good eye for almost a week. Then he suffered an exacerbation of schistosomiasis, which *everybody* at Ndumu has, but when his urine flowed red instead of merely pink, he insisted on having a couple of shots and afterwards lolled about in further imagined incapacitation. (Not that *I* wouldn't have taken the next plane to the Mayo Clinic!)

We were now beginning to receive reports of sick mice from Johannesburg. You will notice that Bruce McIntosh—and Pottie, for that matter—had not joined us in the expedition. *Someone* had to remain at the other end to receive material shipped in from the field and to put it through the virus-isolating routine. And that brings up some matters of logistics which I should already have mentioned.

Dry ice, once again, was at the bottom of all our plans. Bob had had some insulated containers made in Joburg, each accommodating two large blocks, and these were shuttled back and forth throughout our stay in Lumbo (there being no dry-ice factory in Moçambique). Special arrangements had been made in Lourenço Marques so that the customs people allowed our containers to be transferred between Portuguese and South African planes without the delay that inspection would have incurred—in fact, that might have led to loss of viruses if thawing took place. Had there been anything for us to smuggle, we would have had a perfect setup. As it was, frozen carbon dioxide came one way and frozen mosquitoes and animal tissues went the other—as innocent cargoes as ever crossed an international border.

Then there was The Box. Dry ice was delivered only once a week, and Bob decided long in advance that we must have a large box, filled with sawdust, in which to store it. The shipping containers were admirable for a day's trip, but dry ice would not last indefinitely there. He had asked Tito to have a local carpenter in Lumbo build a six-foot wooden cube to which the precious ice blocks could be transferred.

Had I been consulted in the matter, I would have concurred with Bob's directive. The point is, I suppose, that hardly anyone has ever seen a six-foot wooden box. After it was finished, an enormous crew of men was needed to deliver it to camp, and when Bob finally laid eyes on it he was almost overcome. I have since figured that the one-inch-thick planks of a six-foot cube occupy eighteen cubic feet of space. Tropical lumber often sinks, giving it a density greater than that of water. Water weighs somewhat more than sixty pounds per cubic foot, so that this behemoth exceeded half a ton! Bob looked no larger than a squirrel burying a walnut as he stowed dry-ice blocks in The Box's dim interior. Needless to say, the ice held up beautifully.

Bruce's reports of sick mice pertained only to inoculations with A. *pembaensis* suspensions. That was not to be entirely unexpected, for most of what we had sent were of that species. However, A. *pembaensis* had never before been branded as a virus carrier, whereas some other species, which we were submitting in fair numbers, had earned that title, though the latter were remaining negative thus far.

It is always a spur to a field team to know that some of its work is paying off. Thus stimulated, one morning I decided to visit the mosquito collectors at the inland site near the huts in order to bring back such mosquitoes as they might already have caught, rather than wait for the entire batch of tubes to swamp me at noon. The Land Rover was idle, so I appropriated it and off I went along the sandy road. I found the boxes already filled, and the boys were glad to seize this chance to get back to camp early. I loaded them in the rear (which had solid doors, so that they rode as if in a vault). A slight incline led from the road to the camp area, and I had to "gun" the engine to reach that hard, level surface.

Directly in front of me I observed David leaning over his table, measuring a defunct bush baby prior to handing it over to his assistant for skinning. I braked the van, but it kept right on rolling. Something was alarmingly wrong with it, and without time to think of anything else, I turned aside just in time and rammed into the trunk of a tree—hard enough to indent the heavy front bumper. All sorts of thumps took place behind me. When I opened the rear doors, I saw a jumble of sprawled arms and legs disposed among mosquito tubes strewn all over the place, not to mention a dozen goggle-eyed, pained faces.

Rodrigues came running up. He could not speak English, but he soon made it clear that he had been working on the brakes. I had driven to the collecting site with the brake-fluid cylinder empty, but the sandy road had gripped the tires so that when I *thought* I had put on the brakes, deceleration had come from other forces. Well, I would leave the driving to him henceforth.

Thus on some mornings I had a few unoccupied moments to look about for birds. It seems a waste to me now that I should have spent three weeks at Lumbo with a hand lens rather than binoculars before my eyes most of that time. However, the avifauna was not too greatly different here from that at Ndumu, both regions lying in the tropical corridor to which I have referred. The Emerald-spotted Wood Dove uttered the same lament in both places, and male Scarlet-chested Sunbirds were equally brilliant.

But despite a meager bird list, I chalked up one species that gave me great satisfaction. The *chefe de pôsto*'s compound was situated on a low bluff overlooking Moçambique Bay. At low tide this shallow expanse sometimes presented a broad area of mud flats extending for several hundred yards offshore. A few terns and shore birds came here then to rest or to feed, though there was never a spectacular aggregation such as the terrain looked fit to invite. Yet such as they were, I must identify those birds if possible.

If you want to feel heat at its most telling natural maximum, trudge along a tropical mud flat on a calm, sunny mid-morning. While the actual temperature will be inferior to that in a desert, humidity is at the saturation point, and the effort of progressively raising mud-laden feet soon persuades one that sun and earth have conspired to mount a summit meeting here.

Even more than on the opposite side of Lumbo, within reaches of the lagoon, this oozing, fetid waste had the lack of human litter to recommend it. Such strong smells as it emitted were those that accorded with the life and decay of indigenous algae and animal creations. To those that properly lived there, then, the air must have seemed fresh. And what I at first took to be jetsam, damning man for his untidy ways after all, turned out to be rightly sprawled about also: these were stranded, leathery sea cucumbers, perhaps dying in the sun, or possibly able to endure irradiation between tides—a remarkable feat of survival if they could manage it.

Terek Sandpipers I had already seen near Port Elizabeth in company with Richard Liversidge, but they were no less welcome as reacquaintances at Lumbo. Their behavior is unsandpiperlike in that they are individualists. Even when several consort together, each one goes his own way as if alone, while most other species perform as if regimented. Tereks have a peculiar trick of running very fast with their heads stretched out low in front and then coming to a sudden halt in the normal, more erect position.

At the tip of a muddy spit that extended quite far into the bay —farther than I could go because the muck became too soft—I noticed from time to time a group of three fairly large white birds that I imagined were terns of some kind. On the other hand they might be egrets or herons, for their legs seemed rather long. Shimmering heat waves rising from the flat made exact observation difficult, besides which I was looking at them into the sun and had to contend with considerable glare. The tide had reversed itself by this time, and by good fortune it now disturbed the birds, which took wing and flew straight toward me. I think they had planned to alight where I stood. On seeing me, however, they veered down the strand for a short distance when, changing their minds, they turned and flew past in the opposite direction. They thus crossed in front of me several times before selecting a distant spot to settle on.

One could not have asked for a better look, but still I had no idea where to classify them. Long black legs were confirmed, yet those slim, pointed-winged birds were clearly neither herons nor terns. In my mind I could see their picture on one of Roberts' plates, but I had not given it much attention. Now I had an excuse to leave the mud flat, being nearly boiled alive anyhow, and in the shade of my entomological tent I quickly turned to the pertinent illustration. Crab Plovers. And at the same time, not *really* plovers, for if so, I should have known them in at least that general way.

The birds are regarded as sufficiently unique by professional ornithologists to deserve a private position in the scheme of classifying the sandpiper-plover complex. The single species of Crab Plover therefore occupies not only a genus by itself but also a family without other incumbent relatives. And well it might, for as I continued to read I learned that this is the only known shore bird that nests in burrows rather than laying its eggs directly on the ground. Its range is restricted to "tropical shores bordering the Indian Ocean," and true to its name it feeds mainly on crabs.

Thus the sea at Lumbo had cast up a new family for my list, and my gratitude was boundless. Not quite so enthusiastically I sampled a few other local marine products. While we took breakfast and dinner at the hotel, noon was too busy a time to leave our chores, and the cook-tent then provided us with native-style victuals. Goat or chicken curries and rice were acceptable (though not to be raved about), but the fish and octopus were usually better nibbled than taken in bulk. On many mornings a local African came along with a basket of fresh fish, beautiful tiny creatures that after cooking seemed to be nothing but spines. He traded also in tentacled things, living conch snails and prawns. Squids or octopi, whichever they might be, became tough in the pot and when done yielded no more than a somewhat musty taste that I did not find palatable. Conches had to be drawn from their citadels with a corkscrew and then looked revolting. Prawns, yes— these comported themselves in hot water as all crustaceans do, emerging lovely, red and toothsome. But how much of my reaction was based on custom and prejudice?

That question is of no importance, for recognition of prejudice does not automatically make things taste any better. If you are reluctant to chew and swallow, it does not matter what towering arguments you raise to oppose that attitude. João Branco—be he forever blessed!—must have sensed that our lunches were not always epicurean. One sweltering noontime when we were about to sit down to one of the cook's nastier concoctions, one of João's waiters, having trudged all the way from the hotel, arrived with a tremendous cloth-covered tray. Setting his burden on a camp table, he pulled aside the cloth to reveal the largest rock lobster I had ever seen. It had been perfectly cooked, the meat removed in strips and garnished with mayonnaise and sliced hard-boiled egg, while the lobster's intact crimson shell formed an artistic background for this invitation to glut ourselves.

In 1957, Bob had been impressed by the appearance of other parts of the province besides Lumbo. Viruses must be prevalent almost everywhere, though their types, vectors, vertebrate hosts and so on could differ greatly from place to place. While Lumbo had been selected on the basis of the mysterious antibody found in sera from its human residents, Bob determined to take a couple of blind potshots elsewhere, if for no other purpose than to make simple comparisons with results obtained at Lumbo. But beyond that, here we were in Moçambique with all necessary equipment;

it would be tantalizing not to take advantage of the chance to gamble a bit. Expeditions don't come along every day, and the next one would automatically be directed in a widely different direction in order to provide the maximum of contrast rather than to detect mere local deviations.

Most of the equipment—tents, microscopes, traps, guns, bird and mammal collections and so on—was ferried back across the bay to Moçambique Island and loaded on shipboard for transport to Lourenço Marques. Jacinto and Rodrigues accompanied the consignment, also riding herd to displaced African assistants (including Jay Mbuzi, who was now disgustingly well). The Box, which Bob had thought might come in handy on future expeditions, was left to rot where it stood, though I doubt it had the grace to decay—it must be there yet. David took off for Johannesburg independently while Bob and I set out from Lumbo with a greatly reduced mobile set of gear, the largest items of which were an inevitable shipping container of dry ice and several boxes of mosquito tubes.

At Quelimane, farther south along the coast, we left the plane and were welcomed by the public health doctor from Namacurra, at whose residence we would work next day. The name Quelimane is neither African nor Portuguese but derives (it is said) from an Englishman's having told a Portuguese that the climate here could "kill a man." Namacurra is fifty miles inland from the coast, and you know what that meant to me: Let's see *A. pembaensis* put in an appearance here!

It did not. With aid from an interpreter, I hastily instructed a few boys in the art of catching mosquitoes, and they did very well indeed. Had they brought in only a few specimens, I might have held doubts about the absence of *A. pembaensis*, but 416 mosquitoes without a single stray from the coast was convincing enough for me.

The potshot struck two unexpected marks at Namacurra when a virus called Semliki Forest from Central Africa turned up in a kind of mosquito not hitherto known as a carrier. That discovery of course had to await the results of mouse inoculations and other tests in Johannesburg. At the time in Namacurra I was simply impressed and pleased by the fact that the boys had caught 97 specimens of *Aedes argenteopunctatus*, a really beautiful mosquito with silver spots on its legs and the largest number of this species that I had ever seen gathered together. Abundance of a species and the virus-carrying capacity often seem to be linked, though there are some negative instances to offset such a generalization.

The isolation of viruses in baby mice requires endless surveillance, for if the infants become sick, their mothers will often eat them. You can't pump out a mouse's stomach to salvage anything —indeed maternal gastric juices will destroy arboviruses almost immediately so that a total loss ensues. I think it was Bruce who, on examining an inoculated mouse group one day, saw that the mother was munching on the last hind leg of the last of her litter. In the nick of time he grasped the disappearing tiny foot from which, after it had been emulsified and inoculated into a second litter of babies, he succeeded in establishing a virus.

On our return to Quelimane we learned that our plane would be several hours late. "Fine," said Bob. "I noticed a small coconut grove as we drove in, and Brooke can identify some more mosquitoes there."

He literally forced me down at the side of the road and rounded up a few boys to scurry about in the grass with catching tubes. Three hundred thirty-nine mosquitoes later, when it was time to return to the small airfield, my legs were so cramped that I had to be helped to my feet. *But,* we were again close to the seacoast and I had tallied thirty *A. pembaensis* in the collection. That put a nice cap on everything.

A potshot at Nova Lusitania, across the Busi River from Beira, missed viral targets, though I found more than enough mosquitoes to keep me feverishly at work. The terrain had been completely cut over, cleared, and put into irrigated sugarcane fields. No *A. pembaensis* would venture here, though the river was a tidal one and crab holes must not be far away. Instead, the many weed-grown channels were ideal for larvae of the two common *Taeniorhynchus* species, and a mere half-dozen boys succeeded in inundating me with them.

I worked for some hours after dinner that night, finally falling into bed when the last tube had been labeled, dropped into the dry-ice container and duly noted in the field ledger. As far as I was concerned, that was my last official act as entomologist to the expedition, for tomorrow we would fly to Lourenço Marques and return to a conventional way of life.

However, I had not counted on Bob's Texan spirit. Some people might well call it slave-driving, but in his case the excessive buoyancy had no such underlying intent—he simply had a permanent affliction of enthusiasm. When he shook me early next morning, he was bubbling with ecstasy.

"I had the most marvelous time last night," he said. "I was wandering around the town after dinner and I saw some people sitting out in their yard. They were very friendly and asked me to join them." (I'll bet Bob was the one who asked!) "Soon we began to be pestered by mosquitoes, so I came back to get you. You were already asleep, so I took the tube boxes, and the man and his daughter helped me until we filled every one. Here they are, right alongside your bed. Hurry up and you'll have time to sort them before we have to leave."

Taeniorhynchus, Taeniorhynchus! Of 2,555 mosquitoes taken in Nova Lusitania, 2,326, or 91 percent, were *Taeniorhynchus*. That was more than I wanted to see again during a twenty-four-hour period, at least for the time being.

Moreover, I had another urgent job to do before we caught the plane. Dr. John W. Hampton at the lab was doing some very technical biophysico-chemical work with West Nile virus. For his purposes he needed a special medium in which to suspend virus particles during their ultracentrifugation and somehow he had discovered that the ideal fluid was the so-called "blood" of a large terrestrial snail called *Achatina* (the kind that has now invaded Florida). (I believe that the reason for such a selection was that protein molecules in the blood of *Achatina* were intermediate in size between two "families" of virus particles—an important consideration in sorting the families into pure lots. Anyhow, it was something like that.) This snail, inhabiting the tropical corridor, occurred at Ndumu but in small numbers. We found it most often in the fig grove at Site 18, but its populations were so scanty that sometimes John's work had to be halted until we were lucky enough to discover fresh snails for him.

In roaming within a few blocks of our hotel in Nova Lusitania I observed that the entire town was crawling with *Achatina* snails. They covered tree trunks, fence posts, walls of houses—even sidewalks and streets. Entire gardens were defoliated by them, and it seemed that they were so prolific as to be their own worst enemies, threatening the race with starvation for lack of further suitable plants for them to eat. I ascribed this excessive profusion of snails to land clearing. The urge to put every available square foot of soil into cane cultivation had blotted out whatever enemies the snails might once have had, and now there was no remaining check on their reproduction and survival.

Surely we must take advantage of this godsend for John. Somewhere we located an old burlap bag, and the rest of our journey to Johannesburg was encumbered by that bulging item. I would not trust the sack to the luggage compartment of the planes on which we traveled, lest the snails be crushed or overheated or otherwise injured. They went right with me to my seat, where I placed the burden firmly between my feet. I don't know what the several stewardesses thought of the puddles on the floor after I deplaned.

The ultimate results of our expedition were startling. *Aedes pembaensis* yielded eight strains of a new virus that eventually received the name of Lumbo virus. But as in the case of Germiston virus, it was *not* the one we had sought. Viruses are classified in several groups, if they can be classified at all. The Lumbo antibodies detected in 1957 fell into so-called "Group B" in the arbovirus hierarchy. But Lumbo virus was worked out as a new member of the so-called "California Group." A second new virus, named Mossuril for the district in which Lumbo lies, was isolated from *Culex* mosquitoes and had no detectable relationship to anything else, i.e., it was "ungrouped." We failed to find a responsible Group B virus corresponding to the Group B antibodies of 1957; therefore these remained as mysterious as ever. Probably an even greater conundrum was presented by the facts that less than 10 percent of Lumbo residents showed evidence of Lumbo virus antibodies, whereas tests of blood from a series of engorged *A. pembaensis* showed that the mosquitoes had fed almost exclusively on human beings. Where, then, had they picked up the virus, and why did they not transmit it more often to man? No one could say.

Tito greeted our arrival with plans for a celebration in Lourenço Marques. We must go out on the town, though first he must take us home to meet his wife and children. Along an entire wall I noticed a vast collection of comic books. "You allow your offspring to read all those?" I asked.

Oh, no. These were all Tito's, saved since he had been a boy and, in fact, still being added to. Stacks along the wall were only overflow from a room that was completely filled with them. Everything was fully classified and indexed, so that Tito could put his hands on a file of almost any character you might name. He was particularly strong in the science-fiction department. That collection must have been extraordinarily valuable, particularly in view of some of that early fiction now having become fact.

But on we went to Tito's favorite sidewalk restaurant where one was served either chicken or prawns *pirri pirri,* which meant so hot that no conceivable quantities of beer could put out the fire. We were very gay, and as I looked about at people passing by— most of them Africans—I observed that they all seemed to be gay, too. Indeed, on reflection, I remembered that everyone at Lumbo had been gay. Somehow, in crossing the boundary of South Africa, we had left behind the sense of oppression and frustration that emanated from Africans one saw there (except for our blithe savages at Ndumu).

I discussed that anomaly with several people and finally came to understand it on the basis of one word: citizenship. No matter what misdeeds the Portuguese have been accused of by the rest of the world, they make it plain to all their people that they *belong* to the nation. Moçambique is not a colony. It is part of the homeland just like our states of Alaska and Hawaii. Each man, woman and child is a Portuguese and, even if black, has full rights to improve himself (though the opportunity hardly exists). Interracial marriages are personal matters having nothing to do with government policies or prohibitions as they are in South Africa next door.

In practice the blacks might be as thoroughly dominated in Moçambique as in South Africa. The white *chefe de pôsto* in Lumbo, for example, had authority to order his black policemen to round up thieves and murderers and then to sentence them to fines or flogging without benefit of a trial or recourse to appeal. But he did this with as much justice as it is given us to possess, and there was no discernible resentment against him in the community.

Finally, the black could feel that he was noticed and that someone in authority cared about what he did with himself. Children were encouraged to go to school (even granting that there weren't enough schools to hold them). Somehow everyone was given sufficient attention and discipline to make him take a spark of pride in his status. On the morning when I appeared to give directions to my mosquito boys in Nova Lusitania, the local *chefe de pôsto* had lined them up in an orderly, silent row with a policeman at each end. At the *chefe's* signal, they greeted me in unison with a chanted *"Bom dia!"* Yet the moment the policemen turned them loose, they were at their boyish games. I think they sensed their first-class citizenship.

16

SHOKWE PAN

Jay Mbuzi had to wait in Johannesburg until Pottie and I were ready to make our next sortie to Ndumu. Arrival of the van in Ndumu evoked a rapid gathering of the same crowd that had seen us off two months ago, but Jay would not look at them. His status had now risen until his kinsmen were hardly to be recognized. He made a great show of busying himself with unloading our equipment. At last his wife pushed their little boy forward from the throng. The tot, wearing only a short shirt that failed to reach his navel, held out a tin plate with two dark objects, each about two inches long, on it. This was Jay's welcome home token, and he finally could not resist further. He popped the roast beetles into his mouth and crunched happily. (I ate several of those delicacies later and found them tasty, much like hollowed-out roast chestnuts.)

No matter how often I returned to Ndumu, I always found endless sorts of new creatures almost at the threshold of our quarters. This time, as if to supplement lore concerning insects in one's dietary, Qmba Ngwenya illuminated my mind by pouncing on a black, pot-bellied ant that was crossing the open compound. It had a head and thorax of average size, but the abdomen was swollen to the volume of a small cherry. That pendulous extremity was so distended that the ant could not make normal progress but had to drag itself along cumbersomely. Holding the insect high by its foreparts, Qmba Ngwenya lowered the "cherry" into his mouth and bit it off with as much gusto as if this were truly a fruit.

On my walks near camp for the next week or so I found many such ants, though after that they disappeared completely. These were young, fertilized, egg-laden queens that had flown from an overcrowded ancestral colony, then dropped to the ground and lost their wings, and were now dispersing further to establish new outposts of their own.

This was the riskiest part of their journey, for the lumbering ants would be the easiest of prey to lizards, snakes, small hawks, and other insect-eating birds. Now I added myself to that list, for while they lasted, I used to fill a shirt pocket with fat ants and deliver them to Qmba Ngwenya. His eyes would shine as he consumed them one by one, and I almost brought myself to making the gustatory experiment personally. In the end I argued that my little assistant needed the protein more than I did—or *was* that the true reasoning?

Another creature that fascinated me—there being no question of food in this case—was a large caterpillar. I discovered it by observing its droppings on hard bare earth under a small acacia-like bush near Mamba's marula tree. Following the clue upward, I soon beheld the spiniest caterpillar I had ever seen. It was about four inches long and reposed on a narrow twig, with every part of its body armored in all directions. Tufts of heavy black spines protruded from its back at every segment, while longer but more slender brown ones extended laterally and ventrally so that no part of the caterpillar's actual body could be seen. This, I felt certain at once, must be a relative of the familiar Io caterpillar in Pennsylvania, though that one is not so thoroughly concealed behind its protective armament. Even without such presumptive knowledge, I would automatically have recognized danger in the organism now before me, for those myriads of needles had obviously not been set there for mere ornamentation. If an Io could

sting uncomfortably, this one must have been prepared to up-
grade the experience many times over.

Carefully not coming into contact with any part of the beast, I
cut off its twig and carried it back to camp. Qmba Ngwenya at
once beckoned me to come with him. He led me in another direction, into a thicket of the same acacia-like shrubs which I quickly
found was used routinely as a latrine by the mosquito boys. That
was how Qmba Ngwenya happened to know about the caterpillars here. Whereas I had been delighted to find a solitary one, I
was now within a company of them and was able in a few moments to collect a dozen.

The size of those larvae indicated that they were practically full-
grown and close to the end of their feeding period. Accordingly I
made no attempt to provide them with fresh leaves but simply
tossed them into a hastily built cage of plastic mosquito netting.
Within a day or so they began to spin cocoons. Ios belong to the
great family of silk moths, Saturniidae, although only a few of its
members are commercially useful as sources of silk. *This* one
surely would never qualify in business, for it took pains to incor-
porate its spines in the cocoon. How it managed to do that, I can't
say, but the feat must have needed great patience. When a cocoon
had been completed, it bristled with detached spines almost as
porcupinelike as the antecedent caterpillar. (The moths that
emerged were disappointing in their relatively somber shades, for
many saturniids such as Cecropias, Polyphemus, and Ios, too, are
brilliant. Nevertheless I was gratified to see that a family relationship could be detected even in these subdued relatives. Thus,
though I never learned their exact name, I knew I had placed
them in the correct category.)

Lack of a local insect book in South Africa was a constant irrita-
tion. Whereas I could discriminate among the finest details of
mosquito anatomy and classification, I was forced to lump far
more conspicuous insects under such rudimentary titles as "butterfly," "grasshopper," "beetle," and so on. That did not stop me
from admiring and collecting them, but I chronically wished that

they could say more to me. For example, what was the life history of that strange, nay, repulsive, beetle which often sprang our Sherman box traps that had been set for small rodents? This Goliath was without doubt the most unattractive beetle of my career as a bug watcher. It was spherical in shape, about the size of a large horse chestnut, but of a dirty gray color and very sluggish in its actions. Its only defense was to secrete a nasty thick mucus. Well, that was active defense; from the passive standpoint it could rely on the hardness of its rough "shell." I could not drive the stoutest of our insect pins through its body to prepare mounted specimens but had to resort to hammer and nails for that purpose. But that is all I ever learned about the beast.

My "collection," if I may call it so, evolved in an odd, impromptu way. Sometimes I went out with a butterfly net just for the sport of it, but more often the mosquito boys would come in with specimens that had caught their fancy as being conspicuous or strangely shaped. (It did not take them long to realize that this made "The Quiet One" happy.) I would then pin and spread the insects on a board until they had dried enough to keep their poses, but after that I had no boxes in which to store them. If I were to leave the specimens on a table, they would be eaten at night by multimammate mice and cockroaches. Well, in that case, what about the ceiling?

As time went on, this proved to have been an inspirational choice. The low ceiling of the general lab area was made of sheets of composition board of some sort that accommodated pins easily. Gradually I added one specimen after another, moving them about as they became overcrowded so that I could keep members of various categories together for purposes of comparison and contrast. Visitors who came to learn something of our virus work were always arrested first by the ceiling's collection and would go through it, with heads tilted back, from one end to the other. Butterflies, moths in general, a special area for hawk moths, dragonflies, damselflies, beetles, grasshoppers, crickets, mantises, bees, wasps, flies, cockroaches, walking sticks, true bugs, scorpions, spiders: all were there, without names but presenting a panorama of Ndumu's terrestrial invertebrate bounty.

I settled back into my mosquito routine rather restlessly. It was not that the excitement of our recent expedition failed to wear off, but rather that some of its dissatisfactions remained with me. Those unanswered questions concerning *Aedes pembaensis*, its human-blood meals and Group B antibodies in the population didn't yet make sense, and I could not see how they ever would. Moreover, I began to wonder about viruses at Ndumu as well. It was simple enough to identify a thousand mosquitoes every day and to exult when the lab succeeded in isolating a virus from some of them. After all, that was what we were supposed to do. My colleagues in Johannesburg were just as enthusiastic as I in their share of the proceedings. And at the year's end there was nothing any of us could criticize ourselves for along the lines of not having tried hard enough.

Yet the only thing we had to show for that work, really, was a list: so many viruses isolated from such and such mosquitoes collected at specified times in the indicated places. Compared with last year's list, and with lists for years preceding that, one could see—what? That the lists varied from one another, and that occasionally a new virus was discovered. But there were no explanations for such variations or new discoveries. The lists were revealing only for what they set forth; they did not interpret anything.

Even the new viruses were often irritating in a way. They did inexplicable things, or else they failed to give clearcut reactions in the lab. We had two such Tongaland viruses in the background, so that to speak of only seven from Ndumu was not truly accurate. For example Banzi virus—not so named until 1959—had already been knocking around the lab for a couple of years when I arrived. It had been recovered at Ndumu from the blood of a nine-year-old boy. Later, during my tenure, a second isolation of Banzi was effected, this time from mosquitoes in the Highveld, so that we could see no natural or biological connection between the two events.

Nyamanini virus, named for Nyamanini Pan at Ndumu, came from the blood of a Cattle Egret shot, I believe, by Bruce in 1957. As African bird viruses were then scarcely known, this one was a long time being classified and even ten years later had not been given a formal coming-out party in scientific journals. However, by that time it seemed clear that Nyamanini virus occurred as far away as Egypt and that it was transmitted among heronlike birds by ticks.

It was thus possible to stretch the Ndumu arbovirus register to nine types. But there was no discernible order in any of the records. I began to wonder whether there might not be something inherently more basic, from the biological standpoint, hidden in those credentials. And precisely what were the data, anyhow? I had not given the lists much attention, apart from my mosquito contributions to them. But after all, the mosquitoes and viruses went together, and in conjunction they were facets of a more transcendental affair that we constantly referred to as a cycle, without knowing exactly what the cycle was. To demonstrate a virus in a mosquito had nothing "cyclic" in it at all, any more than a single frame from a motion picture can tell us anything about the plot of the complete film. In short, we simply did not know where viruses kept themselves the rest of the time.

Yet might not the pattern of their appearance and reappearance in mosquitoes point to some regularly attendant phenomena that would lead us at least to a surmise regarding the natural history of viruses during periods of their apparent absence? As I now scanned the virus lists, I wondered why some of them could be detected only every other year, or every two or three years. Where were they when the lists remained blank? Why were some years better for viruses in general, that is, for several kinds of virus, when numerous strains of each might be isolated, while other years were almost barren? And when viruses did appear, why would several kinds do so practically simultaneously, even occasionally in one restricted collecting site but not in others?

I put together all the information I could gather, to see what phenomena recurred in rhythm to viral outbreaks, or perhaps immediately preceded them. We had good records of temperature and rainfall in the Ndumu region, going back for two decades. As far as possible I read also what other people had thought on this subject elsewhere, in case their conclusions might brighten my own ruminations. Indeed, conclusions weren't in it at all—hypotheses were fairly common, but no one had come out with flat statements (except for candid admissions of ignorance).

In Brazil, it seemed that yellow fever virus might move about in circles, advancing ever into young populations of non-immune monkeys and leaving decimation in its wake. By the time it completed a circuit, say in seven or eight years, the area to which it returned was always repopulated by a new generation of non-immune animals. That would be an efficient system, but nobody could say that things really happened in such a fashion. I did not like the idea very much, because it was too perilous for the survival of yellow fever virus. If, in its spread, it should reach an area devoid of monkeys, it would die out at once, like a forest fire deprived of further tinder. Once extinguished, it could never recur. Dependence on means for perpetuation as tenuous as that, I felt, must lead the virus inevitably to "miss" at some time. Yet yellow fever, occasionally not in evidence anywhere, always cropped up again. I suspected that it must have other means for preserving itself.

Then there was a school of scientists who proposed that some arboviruses may pass long periods within vertebrates in a "cryptic" phase, deeply hidden in various tissues. By definition, the viruses would not be demonstrable by known methods (mouse inoculation of organ suspensions, for example) during this stage of eclipse. Under certain conditions (that could not be defined) they would emerge and once again be available to biting mosquitoes. I was fascinated by the logic of that proposition, for you obviously can never demonstrate an undemonstrable virus. The position of these theoreticians was unassailable in the laboratory —their negative inoculation experiments were ironically positive in being barren—though I imagine Socrates might have asked them some blistering questions.

When I pieced everything together—local climate, historical records of virus activity, hypotheses naïve and arcane—I found that I had drawn a pretty picture of a zero. Well, then, let's forget viruses for a moment and look at mosquitoes. What, if anything, did they have in common that might be a basis for their periodic dabbling in human and veterinary medicine? Nature, when we understand her ways, is almost invariably found to operate in a manner that is not only simple but that also provides a considerable margin of safety for the survival and welfare of her creations. Mosquitoes are evanescent creatures. The chances that a given mosquito, alive today, will be alive tomorrow are roughly 50 percent. If it is carrying viruses, those pathogens will die with it. That's not very good insurance for viruses—in fact, it is obviously very bad insurance. One hardly expects nature to be content with such flimsy, ephemeral vehicles for infectious particles that exhibit all the traits of immortality common to the races of other organisms.

Water, I decided. Regardless of mosquitoes' divergent forms of behavior as flying adults—whether they hunted at ground level or in the treetops, or sought the blood of amphibians, reptiles, birds or mammals—they all began their careers as aquatic larvae! And, as biologists agree, the water environment is a much more stable one than terrestrial milieus, with markedly less sudden changes of temperature, chemical composition, plant and animal communities, and so on. Was it not possible that the true home of viruses had foundations within the bodies of creatures—or even of vegetable forms—that lived permanently and sensibly in benign realms beneath the surface of fertile lakes and ponds? If so, might not the occasional infection of a mosquito larva (that later emerged as an infected mosquito) be purely fortuitous? In that event, the virus could follow its subaquatic destiny in a perfectly regular fashion, with no thought of terrestrial translocation. Such sporadic ventures in mosquitoes would then not be at all remarkable in their irregularity, since they had not been scheduled; indeed, one would expect their exhibitions to be capricious rather than consistently periodic. And, for that matter, viruses might have other means of spilling onto land, for mosquitoes are not the only creatures that enjoy their beginnings in ponds but forsake such safe havens later.

Perhaps this was only a crackpot idea, but then, many notions, such as the world's being round, were so considered before being proved true. In fact, as I hammered away at the proposition during the rest of my stay in Africa, I became known as a crackpot in that department. Here I ran into a most annoying accusation from some people who should have known better. They failed to distinguish between hypothesis and conviction. Somehow or other they came to the conclusion that I was already satisfied that viruses lived in sloughs and was now doing everything in my power to prove that I was right. Granted it may have looked that way—I only insist that no hypothesis can be proved if all you ever do is sit around and talk about it. But accepting the fact in advance of its discovery was definitely not in my book of errors.

Fortunately my colleagues did not reside in the camp of the critics. Not that they were sold on the idea—they simply maintained open minds, as everyone should; but, far more significantly, they were willing to divert a limited (though, under the circumstances, generous) number of precious mouse families my way, so that I could run actual tests of assorted aquatic organisms for the presence of viruses. Had they been obdurate in that regard (and well they might, for every mouse group was desperately needed for more conventional tests), I might as well have wagged my tail as my tongue.

But where was one to look first? What sorts of things should one collect? The Usutu River (which Portuguese on the opposite bank called the Maputa) flowed past Ndumu every day of the year, maintaining appreciable volume even in the dry season. Fish and other animal life inhabited it at all times, not (as I thought at first) being continually swept downstream, but constantly swimming against the current so that they remained more or less in the same place, thus constituting a permanent or almost sedentary population. However, that environment still seemed a bit too dynamic to me—I should like to play in a less turbulent medium. Viruses in the river might, after all, be Portuguese, or they might come from upstream in Swaziland, whereas I wanted some assurance that they were bona fide residents rather than tourists.

At the other extreme I could find innumerable tiny streams and puddles that burgeoned with living activity after each heavy rain. Some of the inhabitants had obviously arrived by air, either in independent flight (as in the case of many water beetles and bugs) or as ovarian cargo in the abdomens of maternal dragonflies, midges and the like. But snails, seen crawling on mud below the water's surface, must have been there all the time, deeply buried in the hard-caked substrate since the last occasion on which the depressions had been filled. Other types of the microfauna had likewise slept through an antecedent dry period, though in an encysted state, and now broke out of resistant shells like so much popcorn. Yet if the Usutu seemed in its way to offer too much, these small rivulets and hand basins impressed me as inadequate for maintaining the Great Scheme I had in mind. I must neglect neither the river nor the brook nor the transitory pool, but above all I must select a fairly large and reasonably stable arena, where life and love could flourish without turbulent distraction or postponement every time the sun became a bit strong and dried up the microcosm.

Until now Site 24 had been little more than a number to me, though I rather liked the associations it held because we had recovered a few viruses there and it occasionally yielded interesting mosquitoes—indeed my *Aedes aurovenatus* was one of them. The Zulus called it Shokwe Pan. About two miles from the NRC camp, it was a rather long, tortuous hollow between negligible banks on either side—obviously the site of a former stream bed. However, at its lower end, close to the Usutu, an earthen embankment had been thrown up in order (one would think) to form a dam. As I was soon to discover, the stream never had sufficient flow to fill this catchment. After very heavy rains, one could see that Shokwe's level had been only slightly elevated. That meant principally that it now covered a greater surface, for the pan was almost flat and the difference between its lower and upper reaches —at least a quarter of a mile apart—could not have been much more than a foot or two. The only effective way to fill Shokwe was from below. Though it did not happen every year, whenever Swaziland experienced cloudbursts, the Usutu would suddenly come into flood. Overnight most of Ndumu was under water, and pans all along its course temporarily became lost in a general inundation of the entire terrain. When waters equally promptly receded from higher ground, each pan retained a brimming outline, its replenished contents promising good things for man and beast until the next rainy season or, if need be, the one after that.

Even so, Shokwe was a bit larger than I would have chosen if I could make a pan to order. I suppose I would not really be satisfied ever, but that was because the challenge seemed so formidable even when I stood alongside a mere puddle with my dip net. But dipping now became my vocation and the lab soon found itself in receipt of creatures the like of which had never before been injected into mice. The tolerance of my colleagues now proved to be truly of the self-sacrificial variety, for the more aquatic "critters" I collected and studied, the fewer mosquitoes I had time to identify. Thus our basic program was really cut back quite a bit as I amused myself in speculation.

I came to love Shokwe Pan—and the walk to it—in the way that well-known landscapes always appeal to me. I sometimes wonder how it can be that one never sees everything along a route when going over it for the first time. The trees and rocks are all there, and one must perceive them. Yet it is only after several trips that individual objects begin to stand out as different from one another. On the fiftieth or the thousandth viewing, they continue to show themselves in new ways, and I conclude that their versatility is infinite.

Immediately below camp a cowpath led through dense thorn scrub, soon opening up on a slight hillock where a young sausage tree flourished. It was already old enough to bear several large fruits of an ugly, dull flesh-color that hung from stems over a foot long. As far as I know, they were of no use to anything but themselves in the reproduction of their kind. Antecedent panicles of gorgeous blossoms were as attractive as the sausages were repulsive, a contrast in traits that immediately put that tree on my list of favorites from both standpoints. Nor was I less attracted to it when Pottie told me that rustic Afrikaners (if not more sophisticated ones as well) call the sausage tree "buffelbal" in reference to the fruit's resemblance to a bull's dangling scrotum.

Slightly farther on, a White-breasted Sunbird persisted in trying to manage a successful nest in a bush occupied by a tremendously untidy spiderweb. The spider herself was nondescript to begin with: one would think her as safely to be ignored as any of her kind. However, to improve her concealment, she festooned her web with every sort of debris she could lay her claws on—dry twigs, snippets of dead leaves and grasses, silk-encased shards of former insect prey—and as the result of such extremist activity she succeeded in making the web conspicuous after all. Her snare was not of the attractive orb or spoked-wheel variety but more on the order of a shambling cobweb. Probably by design, most of her litter and junk came to occupy the central portion of this dilapidated home, and somewhere within that wreckage the mistress sat.

It was therefore amazing to see that the female sunbird (also by design?) fabricated an adjacent nest which faithfully copied the spider's squalid quarters. This was a penduline structure with a small hole in one side leading to the downy interior. But its outer surface was cluttered with identical scraps of leaves, sticks and grass, even including plant fibers resembling cobwebs, and possibly a bit of actual webbing as well. Anyone passing by would never have thought, "Behold! I see two sunbird's nests." Rather, he might think, "That spider has recently moved a couple of feet from her old station, for there is her former niche beside the new one."

Yet I was not the only one who saw through this duplicity. Whether it was one of the mosquito catchers or a snake, mongoose or other predator, I found the nest torn open one day and the two eggs broken. Meanwhile the true spider home had not been molested, so that the raid represented perceptive discrimination on the part of the marauder. The bird soon built a second nest, close to the original site and again practically touching the spider web, but this one suffered similar destruction. I did not lose faith in the effectiveness of the sunbird's camouflaged architecture: it must often work successfully. The trouble here was that the bush grew almost directly at the center of the path, where the nest would be much more subject to detection than if the spider had chosen a site more remote from a traffic lane which disclosed, on mornings after a rain, the footprints of every quadruped and biped native to the region.

Shokwe was a goal difficult to attain because of all the lures that encouraged me to dawdle on the way. I now tried to blind myself and press toward the pan with only passing notice of the variety of euphorbias that grew interspersed among the acacias, some of them standing independently like great inverted chandeliers, others twining with pencil-thick green stems to form massive crowns on overburdened supporting bushes. I came to Site 16, where we most often caught single-striped mice and were plagued with those spherical cannonball beetles in the traps. Beyond that the cowpath opened on a tree-rimmed grassy savannah, a quarter of a mile in breadth, where two male Pin-tailed Whydahs had staked conflicting claims. Water sometimes collected near one edge of the savannah and temporarily attracted a few Cattle Egrets and Openbill Storks.

Then, after a further stretch of thorny scrub (where I saw my first Broadbill—a new bird family!—and Pink-throated Twin-spots), I arrived at the true beginning of the pan, at its upper end. This was still far from any sight of water, but indicators of its proximity were there. A grove of fever trees—rather puny at the perimeter, but increasingly grand as I passed beneath them—was the signal that one would soon wet one's feet. The name of these trees is rather unusual in the fact that it denotes not a cure for fever but gives warning of fever to be incurred in the trees' domain. Many plants, such as wormwood and others with medicinal properties, are named for their positive values—they point toward health rather than away from it. The fever tree, a member of the acacia group, requires an unusually moist substrate. Consequently wherever it is seen, one may be entering favorable mosquito territory and hence malaria is likely to be prevalent. The natives have known this for centuries and apparently avoid setting up their kraals close to such groves, though they do not hesitate to drive their cattle into them or to traverse their avenues confidently by day when most anopheline mosquitoes are at rest.

They were lovely trees, with strongly ridged trunks from which papery strips of bark flaked constantly. They lifted trim spreading branches that bore characteristically featherlike acacia foliage. Yet their appearance of sturdiness was somewhat deceptive. One could easily press one's thumbnail into the bark, and a pocket knife under only slight pressure cut deeply into the blood-red cambium. Perhaps those traits were somehow associated with the fever legend, for it seemed as if the trees must be internally plethoric or bilious despite their outward freshness and greenery.

I emerged from the grove onto a scene that in some years would have appeared as a lake, fringed on the right with a dense low forest of broad-leafed evergreen trees and on the left by a swampy bed of reeds. I never saw it thus, for a series of inadequate rainy seasons had followed a number of abnormally wet ones, and in my experience the pan never became more than half filled. The forest and the reeds were there, of course, but they were separated from the water's edge practically all the way around by a broad, flat, muddy apron that supported clumps of rushes in a few places but in most of its expanse was an overscribbled tablet of cattle hoofprints.

This impoverished condition, from the aquatic standpoint, was actually an asset to bird-watching. At the lower end of the pan, beyond an elbow midway along its course, sufficient water continued to stand to attract flocks of White-faced Ducks, and I often suspected that White Pelicans soaring over the NRC camp in the direction of Shokwe settled there, though I was never successful in surprising any as I rounded the point. But various long-legged birds found the flats and shallows exactly to their liking. Whereas I might have seen herons of several species in any case, I believe there were greater numbers of them now because low water had concentrated the fish population; and as for sandpipers and plovers that probed for invertebrate life in soft mud near the edges, these would surely have been hunting elsewhere but for exposure of those mucky fields.

Probably African Jaçanas (or lily-trotters, as South Africans call them) were the most discomfited by lack of flood levels, for the partially submerged vegetation over which they usually run was now high and dry. However, Shokwe was their residence. They were undoubtedly still nesting in the stranded reeds, and to forage they were forced to cross open flats distinctly foreign to their taste. Whenever I appeared, they flew back to their shelter in alarm, affording me repeated glimpses of their reddish body plumage and long trailing legs.

Many other birds either congregated here because water was at hand or else merely passed by on missions having to do with different environs. But they could all be splendidly seen, and, since most of them were large, binoculars were frequently superfluous. "Wintering" European White Storks often descended for a grasshopper hunt, while Fish Eagles (relatives of our national bird) joined Giant Kingfishers in dividing the pan's largess. Flocks of Spoonbills occasionally drifted in for a few days' sojourn. Hadedas, strange ibislike birds, appeared in ones or twos from time to time, invariably taking flight when they saw me, during takeoff and long after departure uttering the strident call that gives them their name. I once saw a Long-crested Eagle perched in the top of a fever tree, and often soaring above me would be the stub-tailed form of a Bataleur, one of the most magnificent birds of prey in existence. The evergreen forest might on some occasions be frequented by Crowned Hornbills, large black-and-white birds with long, curved red beaks surmounted by an ornamental casque of similar color.

And so it went, as if my perambulations were no more than bird walks. I have a very particular reason for dwelling on the birds, and that is because they are to me a sort of yardstick for measuring the environment. Perhaps it is that I first became aware of the world through birds (though insects and spiders obtruded almost as quickly). But somehow I feel able to sense what kind of place I am in if I first become acquainted with the birds. Mention a Field Sparrow to me, and at once I am aware of an imaginary scene where, if I were asked to name some of its other components, I would place meadows, fritillaries, fence posts, perhaps a wild cherry tree, and certainly a row of sassafras saplings. A cardinal would conjure up a quite different vision, killdeers another, and so on. Thus familiarity with bird life at Shokwe Pan prepared me, in a way that I can not explain more fully than I have already done, to consider local dragonflies and snails with greater appreciation —perhaps even insight—than would have been possible without avian indoctrination.

All the same, I had a conscience that troubled me on those ventures, even when I returned laden with the wriggling creatures I had set out to capture. Unfortunately this was not the kind of enterprise I could abandon to a hastily trained native collector, for I did not know at first what there was to collect anyhow. Thus I must do all of it myself, and that put a large dent in mosquito work, as I have already stated. My conscience urged me to find ways of saving time, and at last I decided that the delightful walks to and from Shokwe must somehow be abbreviated. Pottie went there in the van with the mosquito boys once a week, and then I could accompany them, but at other times the vehicle was in demand elsewhere. Thus I finally settled on the idea of a bicycle.

When Bob first heard my proposal, he thought I must be teasing, for the idea sounded utterly impractical to him, but I was able to convince him that I meant it. My strongest argument (a truthful one, by the way) was that many of the aquatic organisms succumbed in their bottles from becoming overheated on the long walk back to camp, whereas on a bicycle I could get them there in better condition. At last he concurred and said he would send Pottie downtown to get one. However, what color would I prefer? That had not occurred to me, for color did not seem to have any bearing on viral problems. I said I did not care, so he went in to Joburg with Pottie and they picked out a machine dappled in large black and yellow splotches like a giraffe.

The bicycle was neither the perfect time saver nor the essential specimen conserver I had expected it to be. During a short period of the year, close to the end of the dry season, the path to Shokwe was worn smooth and I could make good, comfortable speed. With the rains, however, everything turned to goo and I had to get off and push, making slower progress than if I had been unencumbered. Then, as the mud became less liquid and began to take on a consistency of modeling clay, the hooves of cattle pressed it into millions of individual pits with sharp edges. These, drying later, formed a corrugated grid over which bicycle wheels bounced so violently that if I had worn false teeth, they must surely have jumped out. Specimens in jars and bottles were jostled and sloshed correspondingly, to their greater damage than if I had walked. Thus the bicycle occasionally languished at the NRC camp, and if Bob happened to come down at such times, he felt confirmed in his first reaction that the whole idea was no more than a not-very-funny joke on my part, for the budget could not afford pranks.

A formidable difficulty attending aquatic collecting was that I could not simply take off my field shoes, roll up my trouser legs and wade into the pan with my nets and dippers. Jay Mbuzi, it will be remembered, had an acute episode of his chronic schistosomiasis at Lumbo, and all our mosquito boys were quite used to the idea of voiding pink urine. A permanent and snail-inhabited

body of water such as Shokwe Pan was a perfect place for intermediate stages of blood flukes to develop. Infected human beings
who urinated here discharged eggs of the flukes, which quickly
hatched and invaded the bodies of suitable snails. Eventually the
snails emitted clouds of fork-tailed cercariae, larval forms that
were now ready to burrow through human skin and develop into
adult worms in blood vessels close to the bladder. I had no desire
to interpose myself in the local cycle. Indeed, I was so terrified of
becoming infected that I not only wore unwieldy boots but also
hastened to dry my hands with a handkerchief if so little as a drop
of water splashed on them.

But wasn't this an excellent place to begin looking for hidden
viruses? Suppose that snails were the basic aquatic hosts for them.
Larval schistosomes could then easily become infected while multiplying in their molluscan hosts. Later, when cercariae emerged
and made their way through the skin of terrestrial creatures coming to the water to bathe or drink, viruses would be passed on to
vertebrates, and mosquitoes could pick them up during the subsequent brief viremic period. No matter what kinds of aquatic creature I considered, there seemed always to be some simple way for
viruses to use them first as hosts and second as vehicles, either
directly or in a secondary manner, for gaining dry land.

Of course there are many kinds of schistosomes, and each is
believed to be extremely choosy about the species of both snail
and vertebrate that it will infect. I did not know what snails to
suspect at Ndumu from the human standpoint, but that did not
actually matter. Schistosomes infecting birds or antelopes could
be just as effective in bringing viruses within range of mosquitoes
as those partial to Zulus.

I did not know beforehand whether or not snails in Shokwe Pan
would be the right sorts to support schistosome cycles of any kind.
That, however, seemed to be almost a certainty on which to rely,
for the numbers of cattle, human beings, wild quadrupeds (including occasional hippopotami that came in at night from the
Usutu) and birds that discharged wastes into the pan provided

such an inviting setup for a cycle that nature could hardly ignore it. The rectal brand of schistosomiasis existed here as well as the urinary one, so that opportunities for the propagation of blood flukes were virtually without limit. I should not fail to mention the presence of several other kinds of fluke—those affecting the livers of ruminants, for example—possessing different means of effecting entry into vertebrate hosts. It made no difference what brand they were, as long as the local snails were habitable by them.

The first test I made with snails gave a positive result immediately. I gathered a handful of them at Shokwe's edge with no trouble at all. They were unremarkable-looking, ordinary, blackish-brown cones about three-eighths of an inch long. At the NRC camp I carefully washed mud off the shells and then put each one into a small test tube containing clear water. Several hours later Pottie called me from my mosquitoes to look at them. The water in most of the tubes had become practically milky from the clouds of microscopic cercariae that the snails had shed. I shivered anew with thoughts of my threatened welfare, but simultaneously I exulted at the sight of these hundreds of thousands of potential virus vectors.

Of course one could think of other ways that cercariae might serve the ends of my hypothesis. Obviously they could not all infect vertebrate hosts, for if they did the hosts would soon be overwhelmed. Indeed, their lives were very brief and they were unable to swim more than a few feet from where they entered the external world, so most of them—much more than 99 percent— would never achieve that critical encounter with skin of the requisite host, to which the perpetuation of their species was bound. So what happened to the losers? You can't answer, "Nothing," for each one was, after all, a life, just as eager to fulfill the cycle as its siblings. They were all composed of organic materials which had been derived from Shokwe's nutrient medium, and nature's economy demanded that their substance be returned to the brew if it had not been siphoned away by terrestrial intruders. God may find it easy to keep track of the fall of each sparrow, but His

infinite powers must be taxed to the limit when it comes to cercariae. Yet the individual cercaria must meet its own unique end
somehow, an event as filled with historical significance as if it
were engraved on a stone monument to commemorate each one.

Many are likely to die spontaneously, or in other words of starvation, for they are unable to feed, being adapted to only a parasitic existence and carrying very scanty supplies of stored energy
in their minute bodies. Disintegration of such minuscule carcasses
would be hastened by action of bacteria and fungi, and one sup-
poses that if viruses are present, these too now find extinction. But
the waters are infested with a horde of predators as well as scav-
engers. Cercariae, propelling themselves by wriggling tails, are
snapped up in countless thousands to provide sustenance for
nymphs, naiads and larvae of creatures occupying an adjacent elevation in the food chain or web. Such "meat," taken raw, carries
its viruses into safe territory, for the tissues lining digestive tracts
of predators may now be invaded directly. Thus mayflies, damselflies, or water boatmen may emerge from their underwater baby-
hood as volant virus carriers that need only be snatched in their
turn by frogs or birds to bring infective agents into the vertebrate
arena.

Of course snails and cercariae might not have been in the
scheme at all. I have told an imaginative story, just to show how
many ramifications for dispersal there are, even if you begin at a
single point under water. Indeed there was another line that I
took equally seriously, though it, too, could have had its beginnings in the snail sequence. Let's assume that mosquito larvae can
become infected by viruses, whether by ingesting cercariae or
from other sources in their dietary. If viruses can survive through
the pupal stage, they will be present in mosquitoes that erupt
from aquatic cases into the atmosphere.

But that possibility would include *male* mosquitoes as well as
females. Males, of course, never take blood meals, not possessing
piercing mouth parts and being restricted by anatomical structure
to lapping sugary exudations from plant stems, blossoms, overripe

fruits and so forth. Therefore the detection of a virally infected male mosquito would be as suggestive a proof of the aquatic origin of the invasive particles as the finding of any other kinds of virally infected aquatic creatures. But how much handier it would be to study male mosquitoes than all those unfamiliar creeping, twisting, slimy Shokwe denizens! I had a crew of trained mosquito catchers at hand, and I myself was prepared to identify males as readily and accurately as females. All that was required was that Pottie issue a few simple instructions in Zulu, indicating that The Quiet One (in his wisdom or insanity) now wanted the same little flies with bushy antennae as well as those with slender ones. Formerly I had railed at the boys when they carelessly brought in males. Therefore one might have expected them to resent the new edict. But they were too young or too pure to think deeply about my vagaries, and, besides, the collection of both females and males actually made their job easier, for they now did not have to look as closely at what they were doing.

Here I must mention that I was by no means the first person to conceive of testing wild male mosquitoes for the presence of viruses. However, my predecessors in that enterprise had undertaken their studies as the result of following utterly different lines of thought. One proposal was that infected female mosquitoes might transmit viruses to their eggs and that, in further sequence, the viruses persisted through successive larval and pupal stages into the next mosquito generation, when of course emergent males could be infected quite as naturally as females. Another suggestion had been that even though eggs were never infected (and laboratory experiments supported that negative conclusion), some infected females might die when they returned to water to lay eggs, and their bodies might be eaten by mosquito larvae already present. These, again, would not feed on a selective sexual basis, so that indiscriminate infection of adults might result.

Thus I was, as far as I am aware, original in proposing that viruses came from a non-mosquito aquatic source. At the same time, those investigations of male mosquitoes *had* been made, and except for a single phenomenal report in a Red Chinese journal (which few people took seriously), all emergent insects had been "clean." The most extensive survey, however, had been hampered by the annoying fact that females had been as devoid of viruses as males during the period of the test, so that one could not draw comparative conclusions. I felt that the field was still an open one. Even though male mosquitoes had been cast somewhat in doubt, they could stand further screening, while the rest of the aquatic fauna and flora cried to be inoculated into mice. Shokwe Pan lay before me in the sun, its surface shining with a brightness that obscured nether secrets as securely as if it had been cased in polished steel. One had only one's little net for the attack, but that could be magnified into an inspirational flag.

17
___

THE S. S. *PATRIA*

You might think that this sort of daydreaming was relaxing —the kind of idyll one would indulge in after a day of hard work, when one had tended to the practical realities of one's job. I did not find it so. The day's affairs, taxing though they were, had become so routine that it required no mental strain to pursue them. Very probably there are lots of scientists who deem it easy to contemplate the unknown and to weave all manner of likely fairytales out of unlikely premises, but to me it is a most exhausting practice. When I had finished my tasks and attempted ever and again to decide how all those mosquitoes and viruses fit together in an efficient and functional living pattern, I discovered that my imagination was constantly blocked by what I and everyone else already knew. Since I believed that that was not enough, I tried to sweep all the old notions away, but it required a more disciplined intellect than mine to accomplish that.

However, there is another way to go about such ruminations: put them aside completely for a while. That, too, requires a tremendous effort and may not succeed in its design to refresh, but it works sometimes. Often I have been stuck on a crossword puzzle and have finally given it up. Late at night I may suddenly awaken with one of the obscure words clearly in mind, or else I may glance at the puzzle next day and immediately perceive an answer that should already have been obvious. Why this is so, I have no inkling. But one might as well try to take advantage of it.

My two weeks of annual local leave were due at this time and I decided to abandon theoretical pursuits for an existence of sheer mental blankness, if, indeed, one could carry one's body to a distant place and leave the brain at home. A life at sea is about as satisfyingly empty as anything I know: one eats, one eats again, and then one sleeps. Therefore I took my holiday on the *S. S. Patria.*

First I had to think about birds, for my mind is never so vacant that it is entirely freed of feathers. The *Patria* was a Portuguese passenger-and-freight vessel, plying between Lisbon and the South African provinces, viz., Angola and Moçambique. If I were to catch her in Cape Town, I might be lucky enough to glimpse a penguin, while the chances for albatrosses seemed better (so far as I could glean from my reading) off the eastern coast where one sampled a bit of the Indian Ocean. I had already been disappointed by penguins at Port Elizabeth, and I reasoned further that such swimming and diving birds might be less easy to spot from shipboard than soaring albatrosses. Hence I chose an itinerary that would take me up and down the length of the coast of Moçambique.

I should long ago have learned that it is ridiculous to call nature's shots in advance. No sooner had the ship left Lourenço Marques harbor than I was smitten with a sense that this would be a sterile trip. Instead of clouds of seabirds following in our wake, I saw only a few Grayheaded Gulls that were mildly interested in refuse cast overboard; and a Cape Gannet or two that simply happened to pass by with no notice of our presence at all seemed to presage an era in which birds would disdain mankind. After that we entered the open sea and, apparently, an avian desert. So it went for most of the trip: the waters remained void except when we were approaching shore.

When you are stuck for almost two weeks in a predicament of

that sort, what can you do? My answer was to pray for as many landfalls as possible. Meanwhile I spent my time in the ship's second-class bar and wrote third-class poetry. Better that than dwell on viruses, for in the latter respect, if not the former, I was strictly "on the wagon."

Nevertheless it is salubrious once in a while to look with appreciation at common birds when nothing better comes along to titillate. In truth, how many African forms could *I* regard as "common," since they had all been strange to me only a year ago? I was confusing the fact that they already claimed a place on my list with the notion that I knew something about them. Should I ignore Lesser Flamingos now, simply because Bruce had pointed them out to me at Barberspan a few weeks after my bewildered arrival in Johannesburg?

Lesser Flamingos, for goodness' sake! Wouldn't *any* kind of flamingo be worth a life's study? And here I was scoffing at them in Beira for almost getting underfoot! I doubt that the city fathers, if there were any, realized what a tourist asset they had in those birds. Latins in general are likely to ignore minor wildlife, though Moçambique was alert to the obvious attraction of reserves for big game and had indeed established a quite large one not far from Beira. Some of the *Patria's* first-class passengers, on holiday from Cape Town and Durban, debarked here to see the rhinoceroses and so on, planning to rejoin the ship several days later on its return from Nacala. But slimy flats at dockside, though brilliant with flamingos at low tide, were probably regarded as so much waste space or, more "constructively," as eyesores that must eventually be dredged out of existence if finances would ever permit measures of beautification.

The authorities would be wiser if they should decide to keep the mud flats as a permanent sanctuary, actually building ramps or promenades on their margins for the benefit of sightseeing travelers, camera fans and, of course, naturalists. Admission could be charged, or else concessionaires selling soft drinks and novelties at entrance points might be regarded as extracting sufficient revenue from the public, to the benefit of Beira's economy, to allow flamingos to perform as free attractions. This is far from a fanciful proposal, from the standpoint of both birds and man. It works very well at the Sagiyama heronry in Japan, for example, where a steel tower has been erected from which busloads of schoolchildren, among other spectators, can watch nesting activities of several species of heron and egret. Since no one molests the birds, they seem oblivious of the gawkers. I could not see that they were disturbed by the racket of buses and cars or by the ring of cash registers at ground level whenever a child bought a candy bar. In what has rapidly become a circumferential housing development in Stone Harbor, New Jersey, another heronry similarly flourishes, the herons, egrets and ibises appearing to be unconcerned about the menacing semicircle of binoculars directed at them from an established lookout.

Yet it is not likely that the citizens of Beira will follow such a policy. Sagiyama and Stone Harbor are delightful phenomena, but they are so exceptional as to make no impression on the current exercise of man's control of the environment. Mud flats, salt marshes, second-growth thickets near centers of recreation or industry—all "wastelands" of that sort are regarded greedily by entrepreneurs. And everyone else seems to agree that such areas should be "put to use," rather than remaining "undeveloped." If wet- and brushlands are of no concern to us in their natural state, is it not easy to conclude that they are worthless to the world at large? Is it not, then, criminal to allow them to continue in desolation without our unasked "rescue" measures?

Beira could undoubtedly expand its port facilities by installing concrete bulkheads to exclude the tide and then filling in with concrete behind such barriers. A sluice, open or covered, could be incorporated for carrying away surface runoff from inland sites. Indeed, health authorities would applaud both measures. Getting rid of mud flats would eliminate a fever-breeding area, while the proposed drain would be vital to sanitation. Wrong on both counts! Exposed flats, washed daily by tides and irradiated by the sun, are inimical to fever-producing organisms; and there are far more sanitary and pleasant ways to handle refuse and human wastes than by discharging them into harbors.

Though many species of bird are inconvenienced by man's steady manipulation of the environment in ways that suit his notions of progress, all but a few find it possible to adapt at least partially to these changes. Perhaps the ranges of some kinds are reduced, and populations are consequently forced to shrink. But other species may be plastic enough to alter their normal diets or choice of habitats and therefore to remain where they originally lived while man infiltrates and modifies their domain.

With flamingos that will never be the case. This tribe of birds has become "overspecialized" to the extent that its anatomy now demands feeding sites that are the exact ones our real estate developers most enthusiastically destroy. The Greater Flamingo is in a somewhat advantageous position, for it is able to seine macroscopic particles from the mud—such readily visible animals as small snails, shrimps, insect larvae and the like. Thus it might learn to forage successfully in a variety of marginal aquatic habitats to which it is now foreign. But the lesser species has gone so far as to restrict itself, by the fine structure of its bill, almost entirely to sifting one-celled algae from muddy media. Therefore it cannot survive on rocky shores, sandy beaches, or pebbled lake margins. Moreover, even muddy perimeters will not give it a large enough arena, for the amount of material it must extract for sustenance is understandably large, while the yield of algae per cubic foot of mud must be negligible. The Lesser Flamingo, in its huge flocks, requires acres and acres, nay untold square miles, of open flats for its daily quota of green soup.

Current dimensions of the problem sound astonishing to the uninitiated. A recent estimate places the world's population of all flamingos at six million birds, of which four-and-a-half million are Lesser Flamingos, the latter confined to Africa and western Asia. One would think four-and-a-half million birds a reasonably large breeding stock, as well as living proof that they suffer from no lack of places to feed and rear their young. True, oh so very true! That is what people might have said about Passenger Pigeons little more than a hundred years ago. Without recapitulating all the forces that have been alleged to conspire in extinction of the pigeons, we can dwell now on the fact that the forests on which they depended were cut down both for timber and to create farmland. With the opening of Africa during the next century, man will surely become aware of the value of unoccupied salt pans and flatlands that can be reclaimed for agricultural use, particularly should his population continue to expand. If pigeons were once reckoned in the *billions,* what can be hoped for a mere handful of much more vulnerable—as it seems to me—flamingos?

Through binoculars at their closest focus, it looked to me as if the birds were feeding on algae that were more than unicellular in size. The slippery banks were overgrown with a stringy, filamentous form that could possibly be broken into smaller fragments in the winnowing beaks. On the other hand, those strands, delicate as they appeared to my eye, may have been veritable trees to the feasting flamingos, and perhaps they were simply drawing "trunks" through their bills to glean truly minute plantules from coarse "branches."

Had I personally arranged the *Patria's* itinerary to give me a generous look at Beira's flamingos, I should still not have allocated three days to the project. However, that is what I was granted because of rain. We were supposed to be loading sugar from Nova Lusitania, but most of the time hatches were closed on both the ship and the barges lest the cargo dissolve before being taken to Portugal. Consequently my flamingos were seen mostly as dark or gray silhouettes in the downpour, but I was fully content anyhow. A flamingo, I remembered, is a flamingo is a flamingo.

My cabin, in the second-class section, was on the lowest deck but outside, so that there was a bit of natural light and I could glimpse the sea. Until we reached Beira I spent as little time in the cubicle as possible, for it was overcrowded when all four of us were in it at once. Two of my companions were rough young Portuguese, probably coming to Moçambique as immigrant farmers. The third, from Angola, made me understand after much difficulty that he was an official of the Seventh Day Adventist church on a tour of inspection of several missions maintained for Africans in the interior. He did not leave my company until we reached Nacala, at our farthest point north. Thereafter I had the cabin to myself.

Except for the Seventh Day Adventist, no one in second class spoke a word of English. To some people that might come as a hardship, and I'll confess that at times I felt more solitary than I should have liked to be. But the alternative—had I been in first class, for example—is always too much the other way. There I would inevitably have been drawn into some aspect of social life and could not have pondered with such satisfaction on flamingos. And now ashore—in Nacala—I might have been prevented from spending an uninterrupted day in contemplation of baobab trees.

I had seen a few of those monstrous creations at Lumbo, but of course we were so busy at that time that I could not take in their essence properly. A fully grown baobab must be seen to be believed, and even then it is difficult to trust one's senses. Indeed it is further startling to see a company of baobabs together, for at once each tree takes on such a different likeness from its companions that one wishes to give them individual names instead of lumping them all under the same term.

Baobabs can be described from several points of view besides the objective one, so I'll begin by saying that they look exactly like the kind of eerie vegetables one would expect to find on Mars. Why we are so convinced that Martian creations must be bizarre, horrendous or insane is not for me to say, but my mind has become conditioned by science-fiction stories and I insist that baobabs belong on our sister planet, not on earth.

At the same time they have something very venerable about them. Again, they appear to have been arrested during violently gesticulating conversation which they were forced to interrupt when you came on the scene but will continue as soon as you are safely out of sight.

Lest these reactions sound extravagant, I shall quote Dr. L. E. W. Codd's *Trees and Shrubs of the Kruger National Park.* "The Baobab is a grotesque, deciduous tree with a large, swollen trunk *60 to 70 feet in girth* [italics mine] and 40 to 60 feet high. Although it has such a tremendous main stem, the branches taper rapidly, giving it the appearance of holding aloft a number of crooked, stumpy arms." You see, Dr. Codd resorts to strong language also. Toward the end of his account he refers to ancient specimens, stating that "old trees periodically collapse in a spectacular heap of fibrous pulp." I never saw that happen, and can only say that this, too, sounds Martian.

A further impression that recurred to me (in the department of venerability) had to do with elephants. Baobabs have fairly smooth, grayish boles not unlike pachydermal hides, while their bulk recalls an elephant standing motionless in meditation on a grassy plain. If you could plant a privet hedge on an elephant's back, the resemblance would be complete.

Baobabs stood out advantageously in Nacala's landscape which —I'll have to harp on again—had its own Martian aspect. The harbor was one of the most beautiful I have seen, with a perfect bottleneck leading from the ocean to a calm lagoon. But roundabout, especially on the northern side, the country was dotted with strange rock formations that were neither hills nor boulders. Rather, they were huge rounded monoliths that looked as if they had been formed *in situ* at some remote time and by processes no longer known in these days. Stationary elephants really should have been there to finish out the massive and monumental trio: animal, vegetable and mineral.

However, barking dogs and bereaved Emerald-spotted Wood Doves reminded me that I was still on the continent of Lumbo and Ndumu. I returned to the ship and its second-class bar, content that my brain had been purged of viruses. Beira came and went again. This time a pair of Pink-backed Pelicans joined me on a jetty as watchers of the flamingos. After that there was nothing but an overnight run to Lourenço Marques.

As I was brushing my teeth at the hand basin in my cabin next morning, I enjoyed the view through the porthole. The sea looks far different when you are down at its level from its monotonous appearance on deck on a calm day (which this was). Wavelets drop their guise of weakness and tell you that you would be brave to dare them in a rowboat. The sun shone on their tauntingly frilled white caps and then suddenly picked out—oh, my lord! why was I down here brushing my teeth?—a huge, planing albatross. I swallowed the toothpaste and grabbed binoculars that were lying on an empty bunk. The porthole, partly crusted with salt and misted with spray, impaired my view, but for a few seconds I glimpsed a black-and-white pattern on the under surface of the wings that matched exactly a drawing I had seen in one of my books. Then the bird (which was not following the *Patria*) disappeared behind a wave and failed to show itself again.

I had brought along my Roberts, but equally important was W. B. Alexander's *Birds of the Ocean,* an utter necessity on any sea voyage. Now I was able, within a few moments, to confirm in *both* books that I had seen an immature Black-browed Albatross. In adults the wing linings are white with very narrow black margins, but young birds reverse the proportions of that pattern with margins so broad that only a central strip is white. Furthermore, of the seven species of albatross recorded from the coast of South Africa, the Black-browed was the most likely for me to have run across, so that all the evidence checked out nicely despite the brevity of my observation. And if I felt a bit cheated for not having had a longer time to enjoy the achievement of my trip's goal, I found solace in the reflection that I must be one of the very few ornithologists—perhaps the only one—who has added an entire new family of birds to his life-list while brushing his teeth.

With that terminal episode, something snapped—or clicked—in my mind. Perhaps it was because we were now in sight of Lourenço Marques and I realized that Ndumu was not far beyond the immediate horizon. Anyhow, one of the missing words in the aquatic crossword puzzle suddenly appeared before me. It was

"Entebbe." If one wanted to study the basic natural history of arboviruses, why not have a look at haunts where they really flourished? At Ndumu we were fond of saying that we worked in the tropical corridor leading down from central Africa. But a corridor is at best narrow, and organisms may flow through it only intermittently. Once arrived in the Ndumu region, they had to endure conditions that sometimes departed far from the tropical norm. We had our dry seasons and our nights with near-freezing temperatures. Mosquitoes fluctuated according to those irregularities and viruses showed themselves to us with even greater inconstancy. Would it be possible to gain a sounder point of view of intricate mosquito-virus relationships in environs where they were not always being upset by the climate?

I can't think why Entebbe had not occurred to me before. After all, Ken Smithburn had come to South Africa from there, and the work he and his group had done throughout Uganda during the previous decade had helped put arboviruses into the common language of tropical medicine. Moreover, Sandy Haddow was still in Entebbe, carrying on from where Ken had left the unit. Entebbe sits right on the equator, at the edge of Lake Victoria; its moist climate is almost constant throughout the year; mosquitoes and wild vertebrate hosts abound in the forests; and altogether there could not be a better place for an inquisitive naturalist to sit and take in the sights for a spell.

18

SKUKUZA CAMP

Bob was in favor of my plan at once. So was New York—for naturally I had to have permission from the Rockefeller Foundation. The only snag that came up was in Entebbe itself. Having received approval at my end, I wrote to Sandy, only to be informed by David Gillett that Dr. Haddow was presently on long accumulated leave in Scotland. Therefore I must wait, though they would be delighted to receive me a couple of months hence.

Well, waiting was no hardship, as long as I knew that I would go eventually, and meanwhile there remained myriads of things to occupy both professional and leisure time. Outstanding during this period was a visit from Merida who, tiring of sitting around Swarthmore, decided to take a trip around the world with a stopover in Joburg to see what I was doing.

Naturally I didn't do any of my usual things while she was there. We went to an unprecedented round of parties (my days as a hermit temporarily suspended), and before her departure we threw a party of our own to show our appreciation for all the fun. I showed Merida the shops in downtown Joburg, and of course we saw the mine dances one Sunday morning. There was never any question of taking her to Ndumu, for people don't make that rugged trip for a mere few nights' stay.

There remained one item, on the "must" list for tourists, which I had missed all this time, and I have Merida to thank for my first visit to the Kruger National Park. That fabulous game reserve is so vast that I shall not apologize for citing a few statistics about its size. Lying along the northeastern border of South Africa's Transvaal province, adjacent to Moçambique, it is a relatively narrow rectangular strip of land measuring 220 miles in its north-south axis and with a variable width of about 40 miles. Those dimensions would give it an area of more than 8,000 square miles, making it larger than the state of New Jersey. By comparison, Yellowstone National Park is less than half as big. From another point of view, one may be impressed by the fact that South Africans have set aside almost 2 percent of their entire land area to establish this reserve (besides which they maintain several smaller ones). But what else can you do when you are dealing with large animals up to and including elephants? Two hundred miles may actually represent the minimum space required, if the objective is to give a reserve's denizens a sense of being free. Some sanctuaries, on a smaller scale, become little more than outdoor zoos.

Unless one went on a guided bus tour, one had to make personal arrangements in advance with the park authorities to obtain permission for a visit. Not that they imposed restrictions on who might be admitted; it was simply because the park was limited in the numbers of people that could be accommodated. Here and there several official camps had been built *and fenced in.* Only in these was one allowed to stay overnight, and each was fixed in its supply of dwelling units and the scope of restaurant facilities.

Our arrangements were rather simple. Because of Merida's travel schedule, we did not propose to "do" the entire park, but simply to sample its nearest corner. For my part, I would have felt a bit uneasy to attempt a more formidable safari in my little Renault anyhow. The farthest I had driven it to date was the nearby Hartebeestpoort Dam–Rustenberg area, and I rather dreaded striking out a couple of hundred miles into terra incognita.

My forebodings quickly began to materialize in fact. About

halfway to the park the car had the first flat tire of its career. It took me a long time to get at the spare, for though the salesman had shown me the tricky little secret catch under the bonnet, I had long since forgotten where it was and how it worked. Then, of course, I had to find a place to get the tire repaired, and that took most of the afternoon. Those details were highly important because they prevented our arriving at the lower park entrance until almost five o'clock. There, after the attendant looked up our application and found that it was in order in his file, we were told that we must reach Skukuza Camp by six o'clock or else be locked out for the night. The camp was about an hour's further drive, at the prescribed twenty-five-mile speed limit, and we must not on any account drive faster. What would happen if we missed the deadline, I wanted to know. The answer to that one evaded my dull ears, but I could see that the park entrance office was about to close, too, and I had visions of our spending the night in the car with wild beasts sniffing at the doors.

As fate would have it, we encountered more animals during that hour than in our next two days combined. That may have been because of the time of day, when the heat had slackened, siestas were over, and evening browsing was in order. But we could not stop, however much we wanted to. I kept my eyes on the curving gravel road, in case some behemoth lurked around each bend, while Merida served as lookout on both sides. When she would call "Zebras!" I *did* slow down a bit, but chiefly I kept those clanging gates in mind.

Then, on a rise with a slight twist to the left, I thought I saw some posts planted in the road, as if we were approaching a barrier, though I was sure we had passed no detour sign. Those posts led up and up, and at their summit was the head of a giraffe, looking down at us from another world. Merida and I were both stunned by the utter weirdness and yet perfection of this creature that stood there barring our way, with neither animosity nor trepidation, but with an aloofness that immediately classified us as the intruders. If it had remained rooted there all night—and it had a right to—our problems would have been settled automatically.

Car horns were another taboo in the park, and I had no illusion that the Renault could push its way past those great legs. But apparently the giraffe had only been crossing the road in the first place, for Merida now spotted several others feeding in a grove to the right. This one must have stopped at the sound of our motor to see what was going on, as if it had not met hundreds of cars before. Soon satisfied, it *sailed* off to join its companions. And "sail" is what I mean, for giraffes at a gallop seem to soar through the air. I had noticed this in moving pictures but had always thought that the camera had been speeded up to produce slow-motion sequences. Not so: the animals cover ground in great leaps, the breadth of these producing a false appearance of less than actual speed and consequently a gliding effect. This may be largely owing to one's underestimating the giraffes' size and re-flexly thinking that they are nearer at hand. In spite of their un-gainly shape and the obvious effort that goes into their loping, they are unexpectedly graceful in their "flight."

Skukuza Camp did not appeal to us at all as a resort. The food was inferior, and the rondavel to which we were assigned was tiny, scantily furnished and possessed of only one dim light bulb suspended from the rafters. Even this flickering lamp went off without warning at 10 P.M. when the camp's generator was snuffed for the night. That was a great hardship for Merida, who always reads much later than that before going to sleep. I suppose the people who came here were uniformly early risers, for the best hours for seeing animals are at dawn as well as dusk. Who, then, would need the generator after ten o'clock?

Willy-nilly, therefore, we were on the road bright and early. We learned that it was permissible to stop, though getting out of cars was forbidden. Since at breakfast there had been much talk about lions roundabout, I had no inclination to stroll, but every bird brought my foot to the brake and we made less progress at times than if we *had* walked.

The park is so extensive, and also so botanically varied, that animals are not distributed uniformly throughout its confines. At Skukuza, for instance, elephants were not to be seen. Here the most exciting beast was the lion, and though those great creatures sometimes lay in the middle of the road, exactly as one sees photos of them in travel folders, they did so only irregularly, and many visitors left the park in disappointment because they had seen no lions. The excited gossip this morning had been that a busload of schoolchildren, arriving early, had seen about ten lions, not in the road, but lying just beyond a tall grassy fringe that bordered the highway. Seated as high as they were, the children had been able to look over the fringe and get a splendid view of the animals. Now everybody else, in ordinary low-slung cars, was patrolling the same road, peering through the grass and seeing nothing. If frustration were material, with physical dimensions, there would have been a great mound of it at Skukuza that morning.

Somewhere along the way I stopped to admire a Carmine Bee-eater as it sallied from a perch after insects and displayed its brilliant plumage. The driver of an approaching car, seeing that I had parked and was using binoculars, pulled alongside cautiously and whispered hoarsely, "Where is it? What do you see." I pointed out the spectacular bee-eater to him, whereupon he said, "Oh, *birds*," and drove on in disgust.

It is a pity, though fully understandable, that more isn't done by the authorities to "play up" birds of the park. That untutored and disappointed motorist could have had great pleasure if he had been more broadly oriented. The trouble lies, of course, in birds' being smaller and more mobile than most of the park's mammals. Then, too, there are so many kinds of birds that they become confusing to a novice—although I have already alluded to my own troubles with identification of all the different antelopes. Also, I suppose that prior familiarity conditions most of the spectators. Everyone has heard of lions and giraffes since childhood. Would even one fifth of the tourists gaping at the pyramids or the leaning tower of Pisa be there if they had not read about those wonders in school, or, more likely, if some of their friends hadn't made the trip last year?

The park must be even more of a loss to botanists than to bird watchers, owing to the command that all persons remain in their cars. Entomologists, too, are losers, for people in those categories must be content with what they can find in the fenced camp areas. One must admit that the park is, after all, a reservation for large mammals primarily. But a tantalizing by-product of this fact is that all other forms of life benefit from the absence of shooting, the control of fires, and prohibitions on timber cutting, opening of land for agriculture, invasion of natural pastures by domestic animals, and so forth. Therefore birds, too, are tamer than usual, and some forms that are scarce or persecuted elsewhere can be seen to good advantage here even from an automobile.

Red-billed Oxpeckers, for example, used to be common at Ndumu, I was told, until dipping of the natives' cattle became compulsory in order to rid them of disease-carrying ticks. Pottie had known the birds well as a boy and could recognize their churring calls. But in all the time we spent there, I never found one, nor did he hear any. Now, in Kruger National Park, I was delighted to find them almost abundantly. They were particularly easy to see on giraffes, for although they clambered over all parts of various kinds of game animals in their searches for ticks, they became most evident when canvassing a giraffe's neck. It is said that oxpeckers sometimes produce open wounds by removing ticks and thereafter keep the lesions open or even enlarge them so as to feed on crusted serum or blood at their edges. However, that cannot be a serious problem, for otherwise the animals would surely be constantly engaged in trying to flick their avian vermin-exterminators away, as they do when attacked by biting flies. On the contrary, though one cannot tell that oxpeckers are actually welcomed, they are wholly tolerated and make free to investigate all areas of their hosts' exposed surfaces. These birds, though related to starlings rather than woodpeckers, hitch themselves along hides exactly the way woodpeckers progress on the bark of trees, a feat which seems to me an even more singular adaptation.

Another bird that I did not see elsewhere, though its range includes all the areas in which I worked in South Africa, was the Ground Hornbill. These large black birds, three and a half feet long with white outer flight feathers, look something like vultures, especially during their heavy flapping while rising for short distances. We encountered a group of four in the road, where they might have been feeding on small carcasses *had* they been vultures and had the park's speed limit not precluded road kills.

But I was not obstinately restricting my interest to birds. The point I have tried to make is that they were there to see and enjoy, and when we finally left without having spotted a lion, I did not feel that we had been totally cheated by any means. And of course antelopes came into the experience in splendid numbers. Almost everywhere we drove, early in the day, herds of impalas dashed across the road or disported themselves openly in flanking galleries. These were the smallest species we saw, standing three feet high and measuring about four-and-a-half feet lengthwise without the tail. That may sound fairly large: in fact a dog of that size would be included among the greater breeds. But impalas are so slender that they give an impression of little bulk, and I was actually astonished to learn the dimensions I have quoted when I looked them up.

For a while I kept track of the impala count, but after forgetting to mark them down several times, I finally summarized our two days' tally with a conservatively rounded five hundred. That number scarcely accorded with reports that impalas were lions' chief prey in that area, so there must have been many more of them. If impalas were distributed uniformly everywhere, not just along the roadside, I suppose the carnivores could have been supported, for that would have meant thousands of impalas, whereas lions were probably widely dispersed, a single pride ranging over considerable territory.

Burchell's zebras, of which I registered seventy-five, may have come in for their share of slaughter, too. In that case there need not be as many, since one carcass might feed a family of lions for several days. However, I still found it remarkable to realize that lions probably want to eat some time during every twenty-four hours, just as we do, and that the game was able to keep its reproductive rate high enough to satisfy those appetites and maintain its own numbers at a high level simultaneously. Moreover, I marveled that zebras, impalas and other antelopes appeared to be so carefree and casual about exposing themselves. One would think they might go about cringing or sneaking, or trembling continuously in apprehension of that sudden dash from behind a tree or a rock that meant their end. If we, unnatural prey that we are, were constrained to remain in our cars, weren't impalas in greater risk at every moment? Granted that they were blessed with oblivion and could not dwell constantly on death as we do—well, part of the time anyhow—I continued to feel that there was something amiss in their apparent lack of concern. Nature should have provided them with at least an outward guise of being eternally alert. It was very well for lions to become tame, but *impalas?*

In all we saw fourteen species of mammal, three of which I did not identify fully because the guidebook did not bother about rodents, bats or even mongooses. An animal had to be larger than that to qualify for a paragraph. To see hippopotami, we drove to a special parking area placarded with large signs admonishing us to wait in the queue until a guard arrived to take us to a lookout point. Here he stood near us with his high-powered rifle at the ready while we looked through binoculars at the backs and snouts of four lazy animals spending their day in submerged comfort. At least it gave us a chance to stretch our legs, but otherwise I would recommend the zoo for hippopotamus-watching.

Twelve kudus and about fifty blue wildebeests, three warthogs, two vervet monkeys, and two side-striped jackals got on the list by name, but not for having put on any sort of performance. That leaves sixty Cape chacma baboons and one spotted hyena to talk about.

Baboons are generally of a rather ugly mien, whether in appearance or behavior or both. This species ("chacma" being its Hottentot name) was certainly the opposite of handsome, though

I can say nothing about its personal traits so far as man is concerned. It is said to have been a great trial to early settlers in southern Africa, quickly learning to pillage crops and often escaping vengeance by its cleverness. Nevertheless considerable numbers were ultimately destroyed, and continued persecution has reduced them in many places to bands living in isolation on cliffs and rocky hill ranges useless for agriculture. Pottie's father contended with chacmas as a young farmer in Natal, though in his old age he used to spend the day sitting with them on a crag, insisting that he be alone so that he could hear what they were saying to him.

Near Skukuza we encountered fearless troupes of these baboons at several sites. They might be sitting in the road, on the side of it, or under trees a short distance away, but except for animals that actually blocked the right-of-way, none moved for other than its own purposes: we might as well have been absent as staring at them. They were almost black and I failed to see a single one whose fur appeared sleek or tidy. Perhaps that was how they were supposed to look, but at the rate they were grooming each other at all times, I suspected that lice were rife among them. In addition they seemed scrawny and the babies looked malnourished. That could, of course, have been a false impression, but I felt that what these animals needed were some neighboring Afrikaner farms with rich fields of succulent mealies. The game reserve afforded them sanctuary, but perhaps it was deficient in natural foods for baboons. Or possibly protection had led to overpopulation, so that the area, though favorable, simply could not provide for the numbers of chacmas now living off its limited bounty. If lions could climb trees, perhaps this species of prey would have been more vulnerable and hence make a smaller *but better* show.

Roads in the park were kept in good repair by gangs of African workmen. On our first morning I cursed a large truck filled with laborers, behind which it was my bad luck to find myself. It was going twenty-five miles an hour, so I could not pass it, and of course its superior noise frightened away everything in its path. The men must have been trained to be considerate of visitors, however, or else they had all become animal fans by working in the park, for as the truck approached a slight rise it went slower and slower until, just before the crown, it stopped. The men now waved at me, in silence urging me to pass. I got the idea at once that there was something to see, and sure enough, a lone animal was squatting in the road about a couple of hundred yards ahead. This was our spotted hyena.

Proceeding cautiously in a series of short runs, I was able to get close enough for a perfect view of the animal through binoculars before it rose and moved in leisurely fashion into adjacent thickets. Until it had departed and we were again on our way, I did not tell Merida that we had just been in greater danger than if our quarry had been a lion. At least that is what I had been told; otherwise I cannot claim that the following allegation is true. Hyenas have tremendously powerful jaws, a keen sense of smell and craven appetites. If you happen to have driven over fresh lion feces, which are apt to contain semidigested material fit to nurture hyenas, the nose may lead those jaws to your tires, which may then be bitten open in the notion that they contain marrow of the same delicious quality. The maxim therefore is, "If you see a hyena, don't stop. Drive slowly, so that you can get a good look at it, but keep the car in gear, and if the animal makes a move in your direction, step on the gas."

I *had* driven over something very offensive while we were behind that truck, and I am sure the tires still smelled of it! Our hyena must already have enjoyed breakfast.

19

ENTEBBE

"Are you the one that's interested in birds?" asked Sandy Haddow at the airport. "I know there's both a 'Worth' and a 'Work' in the Rockefeller Foundation, and one of them likes birds, but I can never remember which is which."

"You have undoubtedly heard of Telford Work," I answered, "but it doesn't matter. Tel is famous for his bird movies, and I'm not famous for anything, but I love birds deeply just the same."

"Well, then, it's all right," said Sandy. "I get sick of visitors who don't give a damn about ornithology, and it becomes boring for me to have to look at birds by myself all the time. I'll try to show you every feathered thing you want to see in Entebbe."

I could not imagine what kind of auspices these might be for my great expedition to headquarters of tropical aquatic arboviruses. On the Air France plane from Johannesburg I had tried to condition my mind to total absorption with the hypothesis, but the stewardesses had done their best to drown my good intentions in cocktails, champagne and liqueurs. Thus my rosy outlook on setting foot in Uganda was intensified by Sandy's first words, as indeed his next ones demoralized me further when he suggested we stop by his club before going to the Victoria Lake Hotel. I can't remember when I finally got unpacked and settled, but it was clear that I must be ready early in the morning for our first bird venture. Did he think I was really Dr. Work and had come here to take pictures?

"Early" was not so premature that I could not have breakfast first and then wait at the front of the hotel for a while before Sandy arrived. And what impressive minutes those were! I had not been able to appreciate my whereabouts the night before, since it had been dark when I was at last delivered to my quarters. The hotel, a one-story colonial structure with long wings on either side of the grand porte cochere, stood at the head of a slope —part lawn and part golf course—leading south to the edge of Lake Victoria. There gentle margins of the water belied this being part of a rim that encompassed over twenty-six thousand square miles, an area greater than that of Lake Michigan. Nor could one easily grasp that the panorama lay almost four thousand feet above sea level or that, scarcely more than a stone's throw from shore, the equator cut our world into equal hemispheres.

Yet those facts were as true as the nearly unbelievable birds that surrounded me. Hornbills were here as hornbills *should* be— great black-and-white birds with huge ivory casques surmounting the upper mandible. The ones at Ndumu were all of lesser species, interesting enough as members of their family, but not what one has been led to expect by trips to the zoo. Besides, these hornbills were both common and bold, flying in pairs across the golf course from one patch of trees to another with the nonchalance of crows! That required a bit of pinching to convince me it was not a relic of Air France's hospitality.

I was totally confounded by another fairly large bird, of which several flew past at close range. They had rather unremarkable gray plumage with a white band across their long tails. The wings were somewhat rounded and the powers of flight were anything but vigorous. My problem was that I could not place them in any category at all. Then there were some shiny bluish-black birds with very long, pointed tails on the hotel roof, mingling with a few perched swallows. Ah! at least the swallows spoke their names; I immediately identified them as wintering migrants from Europe.

Sandy's arrival settled everything, including my *mis*identification of the resident Angola Swallow, though it surely looks like its northern cousin. Rüppell's Long-tailed Starlings must be among the elite of their family, while Eastern Gray Plantain-eaters (the large dull birds I had seen) repine at the nether end of theirs. Plantain-eaters are members of the turaco group, which includes many brilliant forms. I haven't emphasized Purple-crested Turacos at Ndumu, simply because I can't give attention to everything, but acquaintance with that gorgeous green-and-red bird was what put me off plantain-eaters as a member of their group—they hardly seem to go together.

"I'm sorry to be a bit late," said Sandy, "but I wanted to go to the lab first to arrange a seminar for you later this morning. We must hear about your work on aquatic viruses, and we'd like to brief you on what we are doing and where our field program is concentrated at present, so that you can tell us what you feel you should do while you are here. Therefore we'll just have a quick look at birds now and leave the serious bit till later."

Sandy took me for a short drive along the lake road, first of all stopping in a grove in the botanical garden from which I had recently seen several pairs of Black-and-white Casqued Hornbills emerge. Amazingly enough, numerous others were still present, peering down at us from high perches with small curiosity and then deciding to ignore us. "If we're lucky, we'll see a Snowy-headed Robin-chat," said my host. "Lovely birds. But if not, you can come back later—it's only a short walk from the hotel and they are always here. They're a bit shy, but you must not miss them."

We went on to a minor indentation of the shore on one side of which was the local slaughterhouse. Sandy was about to pass by when I saw dozens of vultures and insisted that we stop. "They're just the common hooded species," he said. Perhaps so, but a life-lister deserves to be observed with respect, even if it is a vulture.

This was a rather small sort, actually found in South Africa, too, though I had missed it there. The birds sat heavily in a tree, having apparently fed recently.

"Come on, then," said Sandy, and I suppose he was right. Those vultures would have lingered there all day, doing nothing. What they had to teach had been conveyed in the first minute of seeing them. But before he could put the car in gear, two great long-legged birds rose from the edge of a narrow grass-grown drain leading from the slaughter yard. They flapped up, facing each other, and quickly sank back to the ground. Now they danced about in circles with spread wings, only to leap again into the air. They were largely black and white, but with patches of gray and chestnut, and each bore a conspicuous rounded pompom of yellow feathers on the back of its head.

South African Crowned Cranes were obviously old stuff to Sandy just as Hooded Vultures were, but it is more difficult to become blasé about them. In this instance he allowed me ample time to watch them complete their nuptial choreography. During the interval I became conscious of a note that was unfamiliar but in the same breath almost as well-known as the national anthem, a new variation on an old theme. "*Blue*-spotted Wood Dove" (italics mine), said Sandy. When we had located the source of cooing, I simply had to take the doctor's word in the matter, because it would have been necessary to hold the bird in one's hand to tell that its wing spots were metallic blue instead of green. But generic similarity extended to the voice, too, and I wondered what tragedy this tropical cousin was lamenting, if not the same one as my emerald-spotted mourner at Ndumu.

Sandy glanced at his watch. "Sorry," he announced. "That's all we can do now. As soon as we move this seminar out of the way, we'll get at birds in earnest."

First my host showed me through the various laboratories of the East African Virus Research Institute—counterparts in all respects of our setup at ABVRU, with provision for virus isolation, serological testing of blood specimens, an insectary for rearing mosquitoes, and so on. The staff, too, was about the same size as ours. Among them I discovered one investigator who was actually thinking and working very much along the same lines that my own imagination had taken. After my recitation, Dr. J. David Gillett gave an account of a series of experiments designed to disclose whether monkeys might become virally infected by eating infected insects. Moreover, he set up his project in what I considered a most elegant fashion, for he wanted his results to represent something that could conceivably take place in nature rather than to document a sheer laboratory curiosity. For example, other workers have shown that caterpillars of the bee moth, feeding on wax hives, may be infected by inoculation of yellow fever virus, but that seems to me to have nothing to do with anything, since bee moths would never spontaneously cross paths with yellow fever virus. David first went into the forest to learn what kinds of predatory arthropods inhabit trees where monkeys lived. Then he caught some of those insects and spiders and offered them to captive monkeys to see whether they would be eaten or rejected. Once he knew that monkeys liked a particular sort of leggy tidbit, he let those predators feed on mosquitoes containing yellow fever virus. If the virus survived and multiplied in them, he would have a theoretically perfect chain of transmission: mosquito bites viremic host; predaceous arthopod feeds on mosquito; monkey eats predator and becomes viremic. (It had already been shown that monkeys can be infected by feeding directly on virus.) Alas! The viruses died out in the arthropods, adding this experiment to the mountain of negative results that attend arbovirus research. David deserved a better reward than that for having conceived his project so intelligently.

Another suggestive line that the group was currently considering had to do with mosquitoes breeding in tree holes. This, on a microcosmic scale, was almost identical with my visions of Shokwe Pan as a reservoir for viruses in the permanent aquatic fauna. In this case they were thinking about protozoa, unicellular (or acellular) animals that could feast on the body of an infected female mosquito that died after returning to lay her eggs. When the tree hole dried out, the protozoa formed resistant cysts, still carrying the virus they had acquired. With the return of rains the hole filled, mosquito eggs hatched, and larvae fed on the excysted protozoa, in turn carrying the virus through further metamorphosis into the adult flying and biting state.

I was astounded to find this gathering of kindred minds. Having feared that my "confessions" in the seminar would lead to my being laughed out of Uganda, I received the opposite treatment— not only sympathy but enthusiastic encouragement. If the questions of our mutual interest could be narrowed down to forest insects and tree-hole inhabitants, perhaps this was the message I had come to Entebbe to learn. Shokwe Pan was overpowering in its scope. The truth may have lain somewhere under its extensive face, but there were no guideposts to the crucial square inch overlying the exact invertebrate that would yield a telltale virus. Furthermore I had been considering terrestrial forms, such as David's predaceous spiders and insects, to some extent, but had thought of them as secondary agents that might aid in keeping a dry-land cycle going after it had been initiated by other means. It now seemed to me that that might still be the way things worked (had David's experiment been fulfilled), though of course it could also be thought of as eliminating the need for any such fancy biological mechanisms as we proposed when we talked about ponds and tree holes.

But to reduce the search to rotting cavities in branches and buttresses was a most "elegant" concept (I must repeat that adjective). And though I had often looked into such crannies at Ndumu—I could even now picture several very fine ones at Site 18—I had not taken them seriously in a fundamental sense. This I must begin to do as soon as I returned. But now, while in the company of a team that had been led to the present point of view by daily exposure to field conditions in Uganda, I felt it urgent to see some of their own tree holes and to observe their techniques in studying the stagnant contents of those miniature catchments.

Dr. Alexander J. Haddow was a medical man secondarily. His original degree had been taken in entomology, and he had added medicine to his qualifications only when insects led him to the study of viruses. Consequently his orientation remained largely on the side of natural history, with clinical aspects of minor significance in his studies. This was perfect intellectual alignment when it came to investigations of viral natural history, epidemiology, and—for want of a better word—sorcery. What secret magic was locked in the out-of-doors, so complacent in external aspect but seething dynamically where the eye could not reach, ought to be accessible to such a man, if to anyone.

Indeed Sandy had documented mosquitoes' activities in as detailed a fashion as is possible. In various forests near Entebbe he had established mosquito-collecting stations that on stated dates were manned for continuous twenty-four-hour periods. Here he had a number of catchers positioned at intervals from ground level to the canopy. These men caught mosquitoes in individual tubes as they came to bite. Sandy sat on the ground below with a notebook. As each catcher was bitten, he gave a shout so that Sandy could record the exact minute of the event. The tube would then be lowered in a box on a string and Sandy immediately identified the kind of mosquito it contained. Thus after the twenty-four hours—after many such twenty-four hours had been averaged together—he could draw beautiful graphs that showed not only the biting-activity pattern of each species but also its changes in stratification, if any, for some mosquitoes always bit at ground level, others restricted themselves to treetops, while many showed definite tendencies to shift from low positions during the day to higher ones when it became dark and cool. The latter, restless sorts could be regarded as potential agents in the transfer of viruses between canopy- and ground-dwelling vertebrates or, to become personal, between monkeys and men.

"You must visit my mosquito towers," Sandy urged. "I'm going to be tied up, but I've assigned a driver and my head mosquito catcher to show them to you. After you've been up, we can talk further about the significance of all this work."

Been up! The first tower, some twenty miles from Entebbe, in what I believe was named Mpanga Forest, rose one hundred and twenty feet from ground level until it cleared the tallest branches that extended above general canopy level. This structure was made of narrow steel girders and guyed by heavy cables, and was indubitably stable and sound. However, the vertical ladder that led to an observation platform at the top looked immensely uninviting, especially since there were no places to stop and rest along the way. While the dimensions of its rungs were undoubtedly uniform, I saw them as wispy strands of spider web near the summit. All of a sudden I felt that gravitation was an ally to be cultivated where I stood and in no wise to be alienated by traitorous excursions aloft.

"Ah, yes," Sandy had added. "I keep a guest book in a little rainproof cabinet on the platform. Please sign it. You will place your name on a roster that includes practically all the world's famous virologists and entomologists."

I now appealed to my mosquito-catching guide (who fortunately had added a bit of English to his native Swahili). Could he bring the book down for me to sign, or, perhaps, to save him the trouble of making two trips, could he lower it to me on a string? But this paragon had been well trained by Sandy. Although he did not seem in the least astounded by my request, neither was he even slightly moved by it. This must have been a familiar interlude during the tour. Possibly a scientist or two had gotten away with it in the early days, but once Sandy heard of the evasion he had so thundered at his employee that the miserable guide no longer dared enjoy bribery and corruption. He simply pointed at the foot of the tower and said, "You go up."

That list *was* illustrous. I spotted a few names of rather elderly persons who, if their bold signatures had really been written a hundred and twenty feet in heaven, put my shaky pencilship to shame. However, now that I had succeeded in the hardest part of Sandy's assignment, I tried to forget my yet-bounding pulse and enjoy a monkey's-eye view of creation.

"Forests" about Entebbe, especially those near the lake or actually on its borders, were likely to be so small that the term seemed inappropriate if size alone was chosen as the criterion for naming them. Most of the rolling land was either naturally clear or else had been cut over long ago, being now occupied by areas of grass dotted with scrub and low trees. Wherever depressions existed, however, seepages rendered the terrain useless for grazing or habitation and so-called forests rose as islands amid the flanking open slopes.

But viewed from the interior, they deserved every connotation of the seemingly pretentious title. As if to make maximum use of restricted space, trunks sprang up close together and then soared in company in competition for the sky. Vines ascended with them, and if branches already mingled to form a solid canopy, trailing thongs wove the mass further into an aerial mat. Seen from above, the undulant surface of foliage presented a dense appearance except for places where a recently fallen branch had torn a hole in it. There, as well as in the gap produced by the tower, one could glimpse part of the void that extended below, but, the eye failing to probe to lowest limits, one had a sense of being borne on a green iceberg lacking foundations.

Island forests—icebergs—the metaphors were more than apt when I took in the surroundings of this isolated jungle and many similar ones on following days. Particularly when the patches stood near the shores of Lake Victoria, I began to sense that there was something odd about them, as if they were, if not actually anomalous in their settings, somehow unique features or atypical phenomena in a larger landscape which controlled and dominated the character of this region.

Indeed a far more surface-embracing habitat was the papyrus swamp. I don't know how many hundreds of miles—possibly thousands—go into the lake's perimeter, but the number of *square* miles occupied by papyrus is correspondingly immense, for at every indentation of the shoreline that giant sedge flourishes, in larger recesses forming such dense volumes of old and new vegetation that the milieu becomes a special one, neither water nor yet land.

Thus I slowly reverted from an original fascination by forests—
a reaction engendered by Sandy's enthusiasm for revealing the
vertical stratification of mosquitoes in their interiors—to an opinion that the lake and its papyrus swamps might be more elemental
in studies of viruses. Encysted protozoa in desiccated tree holes
were a bit more esoteric than needed to explain viral survival,
when a vast major aquatic reservoir existed constantly near at
hand. Therefore I asked Sandy about mosquitoes of the lake's
margins at my first opportunity.

"That's what everyone thinks at first," he answered. "But there
aren't many mosquitoes out there—at least they are hard to find.
And mainly they are a few species of *Culex*. We haven't found
any viruses in them."

I might have added that it was also harder slavery and much
less entomological fun to toil in the swamp, broiled by the sun or
pelted by rain. Nevertheless, since no one knew where the beginning lay, why choose the most unpleasant places to pry? Fair
enough: adopt the forests, which presented conveniently seques-
tered units instead, and work them to death. Yet I continued to
feel that perhaps the reward here lay wholly in mosquito lore, not
in virology. In short, I was becoming confused rather than en-
lightened.

The tower in Zika Forest (for I continued to observe, no matter
what my negative thoughts) was an older one, not as far from the
lab as the Mpanga structure. Actually it was not precisely a tower
but more of a platform reached by a series of ladders. No guest
book was here to challenge climbers, and the summit had been
built only sixty feet above the ground, still well below canopy
level, so that one could look for some distance in all directions
through the forest's understory. Having learned those statistics at
the lab, I expressed the wish to spend a morning alone on the
platform in order to soak up the atmosphere of the forest to saturation's limit.

I dismissed the driver, asking him to return at noon, and then
took a careful look at the tree. It was a real monster, which is why
they had chosen it. However, in the manner of all trees, it sent out

branches at somewhat of an angle from the vertical, and several of the lengths of laddering had been nailed on the outer—or under—surfaces of these. Well, the hell with timidity: I pretended that the head mosquito catcher was there saying "You go up," so up I went.

Now I was ready to exercise my bird-orientation technique, so that later I could transform it into hunches about what mosquitoes thought of this suspended world. One thing was certain: life existed here in three dimensions, but that should not put me off since I had spent a large part of my boyhood in trees. Things might move up or down as well as forward and sideways without seeming to perform unconventional acts.

Very soon I was pleased to see the approach of a small group of birds with patches of white in their plumage. The overhead canopy shut out much of the sunlight, even though I was not far from the "ceiling." The understory, moreover, was far from unoccupied by leafy branches, so that as the birds drew closer, I had only broken glimpses of them in crepuscular alleys that here and there penetrated near distances. The birds were deliberate in their actions—but why pretend any longer? By this time I had recognized the white patches as belonging to a troupe of slender black monkeys with long red tails. Each one bore a white heart-shaped mark on its nose, and though the animals had light cheeks and underparts as well, it was the nose patch that first commanded notice as a monkey faced an observer in the over-branched shadows.

Later I found that these animals were called red-tailed monkeys and that Sandy had published a most detailed and excellent account of them in the *Proceedings of the Zoological Society of London*. But right now I was not interested in their exact names or other scientific credentials—such as stomach contents. It was more than sufficient unto my cup that they should continue to approach, pausing to rest from time to time, and showing no undue alarm at my presence (of which they must have become aware).

When they had finally passed, remaining extraordinarily silent,

I could almost wonder whether they had not been a mirage or chimera, some product of the jungle's vapors that congealed only in my misguided senses. For now the platform, the tree, this elevated fragment of space halfway between heaven and earth in Zika Forest, took on a further semblance of unreality in the impression that it was not *possible* for such a setting to exist with me in the middle of it. What combination of stars or atoms had it required to transport me from another continent, on vehicles as flimsy as viruses and insect wings, to this wooden prow that cut through drifting exhalations as if bent on transnavigating the forest in time if not in space? I realized that this morning was a peak in my existence for which I had long prepared, without knowing where I would find it. While life still might hold future ascensions, to be fought for and bought with other muscles, a part of me could now be allowed to run down and collapse in neglect: *this* pinnacle could not be capped. Exhalations! or a tropical fever, perhaps?

Strange things could happen here, none of which was more stimulating *and* confusing than o'nyong nyong, one of the topics mentioned at our seminar. This, apparently, was a *new* arbovirus disease, not only in its causal particle but also in its life-history pattern. Sandy attached such importance to it that he assigned me as traveling companion to David Gillett on an investigative trip of several days in eastern Uganda, where outbreaks were occurring. "It even acts differently in baby mice," he said. "If you are looking for evolution of viruses, or hints of any other odd manifestations of viruses in nature, this one may give you more clues than our old conservative standbys. Yellow fever hasn't shocked anybody for ages, and even the Rift Valley and West Nile bugs seem to have become set in their ways."

At that time the new virus, if such it really was, had not been named. Wherever we went, David first asked whether the people had recently experienced a new kind of disease and, if so, what they had called it. He could speak quite a bit of Swahili, and with assistance from our African driver could piece together bits of the various local dialects. And it was always the same: the translation would mean "joint-fixer." Thus we would know that we were still on the track of same virus. Only once did he encounter a variant so divergent as to bear no etymological relation to its predecessors: in that village they called the disease "nylon." Asked why, the people explained that they had never seen either this type of illness or nylon shirts until recently, and since both were new they used the same name for each. David did not consider that candidate seriously, but ultimately chose *o'nyong nyong* from the "joint-fixer" list.

Without question a strange evolutionary upheaval was doing a dance of life before the eyes of peasants and scientists alike. O'nyong nyong must be considered a descendant or offshoot of Chikungunya virus, long known as another type of joint-fixer but conventional in all the usual arbovirus respects, including its adherence to culicine mosquitoes for the invertebrate phase of its life cycle. But o'nyong nyong confounded everyone by multiplying in two infamous malaria vectors, *Anopheles gambiae* and *A. funestus.* This was equivalent in virology to a leopard's changing its spots overnight in the jungle. As Sandy had said, we were on hand at the crucial time and place when a step in the ordinarily slow pace of natural change could actually be witnessed—and not so much a step as it was a leap.

However, that was all too much—and too heady—to be taken in on a short visit. I would need ages to think it over, and even then I might never see more than its surface oddities. This was more than I had asked for: a peep from behind the scenes would have been easier to absorb than a sudden confrontation with klieg lights at stage center. My trip with David progressed in a series of overilluminated vignettes—two convalescent cases of "nylon" seen at Tororo Hospital; one convict in the acute stage bled at Butaleja; onward the next day to Mt. Elgon, with a trout-stocked Suam River on its eastern slope; lost for a time on poorly marked roads and discovery that we had strayed into prohibited Karamoja territory in Northern province, inhabited by primitive Africans, some of whom looked as if narcotized; fifteen *really* wild giraffes sailing across the horizon between Greek River and Mbale. . . .

And then somehow I was back at Entebbe and down to a level of understanding from which I could operate sanely. Sandy affirmed immediately that all the necessaries had been accomplished and that from here on we could concentrate on birds. "Besides," he said, inadvertently looking at his watch, though we still had several days to spare, "there's not much time left."

Obviously I can't mention every one of the birds we saw. Had I been on my own in a strange land, I might have done well to mark up a dozen species and then I would insist on talking about them all. But 157, of which 68 were life-listers, are too many claimants for discussion even at a gathering of ornithologists, so I'll choose only a few of the most elegant ones.

Lest I forget, before I talk further of the birds I want to mention a few insects, too. Sandy could not avoid the lab all the time, and while he was briefly at work I occupied myself with a charge I had received from Hugh Paterson in Joburg. "Equatorial cow-pat flies would be immensely interesting," he had said. "These flies have been studied in temperate regions, and a great volume of literature has been published about them—those that merely feed on cow-pats and those, on the other hand, that breed in them, as well as the succession of forms that inhabit a cow-pat in the course of its destiny from a fresh mass to its disintegration and loss of identity. But in the tropics we don't know anything at all. If you could catch even a few such flies, it would be most useful."

Thus I found myself following a group of cattle near the lab, armed with a butterfly net and cyanide jar. Whenever a cow "performed," I waited at the spot for flies to arrive and then swung the net cautiously in an attempt to catch insects without daubing the fabric (quite a trick, by the way). This was not a profitable venture. Perhaps my interest was not sufficiently keen, but I blamed the hot sun. At least one thing was certain: flies did not arrive in swarms, nor yet in twos or threes—I saw no more than one fly per several cow-pats. That in itself should enlighten Hugh.

Thus engaged one afternoon, I looked up to refresh my eyes

with the vista of Lake Victoria and saw two tawny clouds far over the water, at least two miles distant. They looked something like waterspouts, though the day was so calm that I could not believe in such an interpretation. When I reported the observation later, everyone at once said, "Lake flies!" Apparently the lake breeds inconceivable numbers of chironomid midges that form vast swarms from time to time and occasionally fly to land at night, becoming great nuisances around lights. Could viruses be concerned here? But none of *that* now—here was Sandy to rescue me from my role as cowherd.

Naturally my host had spotted all the highlights long ago, so those were sought out with no loss of time. But since the avifauna was so rich, and because many birds could show up almost anywhere, we enjoyed a double boon—the predictable and the unexpected befalling us simultaneously.

At the edge of a small village not far from Entebbe, Sandy stopped at a tree that was festooned with weavers' nests. Weavers are very common in Africa, both in number and kind, being generally similar in yellow-patched sparrowy coloring and in oversexed behavior of the males. However, this tree held something different. Vieillot's Black Weavers were remarkable for the swains' being jet black and possessing striking yellow eyes that stood out so astonishingly from their background that the birds appeared to be constantly alert, like a hyperthyroid patient with exophthalmos. But their charm lay chiefly in a form of courtship that should appeal to all those who like to romanticize the rose-covered cottage. I would have missed this trait if Sandy had not pointed it out, for one had actually to gaze across a bird's threshold to perceive it.

Often an ordinary male weaver can pass muster with the female when his nest is no more than half finished. Perhaps it is wrong to say that *he* passes muster—it is the nest that a female accepts. But the male goes along with the house for the nonce, and by the time eggs have been fertilized the pair have outworn their fleeting interest in each other. Now the ex-husband begins a new construction.

In the society of Vieillot's namesakes, the females were more demanding or else the males simply took greater pride in their work. Whatever the case, nests were meticulously fashioned in full and then *decorated* before being opened for inspection. These nests, like those of most other kinds of weaver, were retort-shaped, globular objects with a cylindrical neck hanging downward so that birds must enter from below. Vieillot's Black Weavers wove the necks exceptionally short, but they had good reason for such economy. As a finishing touch, they pulled in a nearby green leaf, still living and attached by a petiole to its twig, and sewed it with a few strands of dry grass into the entrance cylinder so that it formed a perfect verdant doorway across which bridal wings would have to pass. A pity the honeymoon's so brief.

Our bonus (as if one were needed) at that tree was the sight of a great Saddle-bill Stork flying overhead. It passed close enough for us to see not only the sharply demarcated black-and-white pattern of the body and wings but even the variegated bill, basally black and bright red in its outer half. But what made the glimpse almost uncanny was that Sandy and I were both peering at the same weaver's nest when the stork flew into range of our binoculars. Such luck, it was now proved, is not restricted to the deserving, for weren't we playing the most flagrant and delicious game of hooky?

Next on the schedule of truancy was a hill overlooking an especially deep indentation of Victoria's margin that encompassed several square miles of swamp so densely grown that the area looked like dry land. "But don't let that fool you," warned Sandy. "You might be able to walk on it for a while, but if it gives way, you simply disappear into the unknown." The waterbuck is a type of African antelope that has become adapted to such terrain, developing so-called "false hoofs" to act like snowshoes by spreading weight on the uncertain surface. We saw some of those animals far out in the swamp, and Sandy said they lived there constantly.

But we had come here to look for Whale-headed Storks, known also as Shoebills. A pair had nested in the marsh as long as local people could remember, which did not mean that the same two birds had lived there all that time but rather that this habitat had remained undisturbed and favorable for generations of Shoebills. Sandy had seen them on several occasions. However, there was some chance involved in obtaining a successful viewing. The storks spent the day motionless, well concealed in the tall vegetation, and might shift their positions slightly only in late afternoon. Oh, they would move about more widely after dark when you couldn't see them, so the only opportunity was just before dusk when the first desultory sorties were made. If you missed then, you would have to come back the next day, and if you missed on too many days you were in danger of finding yourself back in Johannesburg without *Balaeniceps rex* on your list. That's what happened to me, to my great dismay, for this bird has a restricted range in Central Africa and who could guess when I would visit a suitable stork swamp again? Sandy made things worse by shouting, "There it is!" just as my back was turned. A Black-bellied Koorhan—much less important—had come along to claim my view, and by the time I turned around, the stork had dropped back into the anonymous wasteland. "You didn't miss much," said Sandy, trying to be consoling when he saw my disappointment. "It was just a sudden flash of a pair of great gray wings—only a flap or two—and you really couldn't see the bird plainly." He was right: that wasn't much. Yet to me it was the loss of an entire bird family, consisting solely of this one unique species.

I needed some comforting for that one, and nature did her best to comply by producing a Blackheaded Gonolek. What, for goodness sake, could that be? I thought I had a fairly complete idea of the kinds of birds the world supports, but at Entebbe I ran into gonoleks, malimbes, leaf-loves and other fare that showed my vocabulary for the parochial thing it was.

Well—the gonolek was after all only a species of shrike, but I imagine someone felt it needed a special name because it departed so far from the black-and-white pattern of most of its relations. Dress crows in rainbows and you won't be satisfied until you call them birds of paradise. To be sure, the gonolek was black above, but its entire underparts were vermilion, and if that isn't a remarkable shrike I'd like to see one. Sandy heard it calling from the marsh and was thus able to point to a spot where the bird might appear. It ultimately did so with far more grace than the Whale-headed Stork had shown, as it sat high on a swaying papyrus stalk with the afternoon sun reflected from its brilliant breast.

I never asked Sandy about the argument he and Ken had had regarding robins. It was clear to me that ornithology was too holy a field in Sandy's estimation for him to entertain light thoughts of it, and what I regarded as a joke might well have been the reminder of an old wound that had best be forgotten. I think it would have been safer to joke with him about viruses! But bird watching was really the greatest privilege I could have had: Work or Worth, Sandy saw to it that my very last minutes before plane time were still as fully occupied with bird listing as all the former ones had been. We drove through the botanical garden on the way to the strip, just catching the Snowy-headed Robin-chat in the nick of one changing down of gears. Who was the sadder of us, I wonder, at my leaving? Sandy would have a long wait for another playmate as appreciative as I, and I might never again see Lake Victoria, even for the doubtful joy of chasing cow-pat flies.

20

DAYS AND NIGHTS AT NDUMU

The logical and practical thing to have done on returning to Ndumu would have been to get thankfully down to my job of sorting mosquitoes and to forget all about the comings and goings of viruses. For I was now in a worse state of mind than before the excursion to Entebbe. Learning had led only to greater perplexity; 'twere better then to abandon inquiry.

In a sense, that is what I did. Discouragement played no part in it: there was simply so much to do during the last few sessions that I ran like a machine instead of a thoughtful investigator. And in the evenings, when I might have enjoyed a quiet hour or two for reflection, there was always Pottie to be entertained. Not that I resented his intrusion. He was such an excellent field companion in every way that I could not grudge him those after-dinner moments. All day long he kept the camp running properly, supervising Jack and the mosquito boys, and seeing to it that I was not interrupted or disturbed by anyone, *including himself*. Thus, after we took our coffee into the general lab following a hearty final meal of the day, Pottie always settled himself with enormous care and then turned an expectant face toward me with the words, "OK, Doc, now let's shoot the shit."

Well, why not? The air was clean, the natives were friendly, and there was every reason to abhor frenetic pursuits. What *did* we talk about? Women, of course, though I don't remember what we said. And then there were many tales of Pottie's career as a policeman in Durban—or, more accurately, a few such tales told many times.

The first evening a solpugid appeared, Pottie shrieked in mid-sentence and jumped about a mile, for it was heading straight for his left shoe as if intent on a lethal attack. I did not blame him; whatever the creature was, I, too, thought it had offensive notions in mind. All we could really see was a fuzzy brown streak moving rapidly across the floor in a perfectly aimed trajectory. As Pottie lifted his foot, the animal continued without deviating until it hit the wall, when it caromed off and then ran, or almost seemed to flow, at the angle one would have expected if it were a billiard ball.

My immediate thought was that this must be a multimammate mouse gone strangely berserk, because I could think of no other small, swift, furry, brown nocturnal being that might run across our floor. Then, after it had bounced with geometric precision from several other obstacles and come temporarily to rest, I saw that it was too small to be even a young mouse just out of the nest. Besides, it had twice too many legs. With difficulty—also with the greatest caution—I caught the beast in a net and got it into a cyanide jar. Only now could I identify it as one of the "harmless horrible" arachnids that accompany spiders and scorpions as barely mentioned relations in textbooks of medical entomology. Solpugids, known to a few also as sun spiders and wind scorpions, always scare the pants off everybody on first acquaintance, so Pottie and I were not to be sneered at as sissies for reacting excessively to the introduction. Thereafter we enjoyed watching them. Some individuals did not "carom" as expertly as others, but why any of them did it at all confounded me. It seemed a mighty blind method of hunting, if that is what they were doing. It struck me only later that perhaps it was they who were being hunted. Possibly some night thing, even more horrible, had frightened them out of their crannies, and their headlong dashes across the room were impelled by a brand of terror as exquisite in its urgency as Pottie's had been.

Solpugids should make splendid additions to my collection of pinned arthropods on the ceiling, I thought. But they turned out to be the flimsiest of creatures, never drying out properly in the positions I set for them. Instead of maintaining fierce postures, with front legs raised in threat, they simply flopped in a rag-doll gesture of collapse. Since I eventually donated the collection to the game reserve, after the chief ranger had said he would like to display the specimens at headquarters, it was probably better that the solpugids now looked so meek and would not intimidate visitors. Some people would without question rather face lions than a live solpugid.

Our evenings were usually brief, but not because Pottie tired of chattering. The longer lights remained lit, the later I kept leaping up to collect insects that flew against window screens or, if small enough, managed to penetrate their meshes. Among the latter were some tiny *true* bugs that at one time of year were a most unpleasant pest because of the characteristic "stink-bug" smell they emitted when they tangled in your hair or slipped down inside your shirt. I soon tired of collecting those, but larger forms kept me on the run until it seemed to me that lights out and bed were my only recourses to peace and relaxation.

Lights out meant shutting off the generator, so one took a short walk out into the night to stop the flow of petrol to the carburetor. What heaven, then, to "hear" silence and, as one's eyes adjusted to blackness, to see the dark become pricked and shaded with all manner of elemental fires and Stygian backdrops. On clear moonless nights the Coalsack could be found just below the Southern Cross, looking like an area of blackness painted blacker. The Large and Small Clouds of Magellan made me feel like that pioneer, though one had no right to maritime emotions at Ndumu. And when I discovered the tiny constellation Musca, I adopted it as the emblem of this arbovirus station, since "mosquito" is no more than a diminutive of the old Romans' word for "fly."

Sleeping was no problem. Bats in an air space above the ceiling made noises as they ran about on their wrists and ankles, but one became accustomed to the sound, just as one ignored distant tom-toms that sometimes continued throughout the night. An unfamiliar noise would quickly bring one to wakefulness. One midnight, in the dry season, I thought I heard a male toad chorusing in ardent courtship, but finally I ascribed the song to a Moçambique Nightjar. There were singularly few stirrings or rustlings or other extraneous susurrations outside, suggesting stealthy movements of animals or human beings prowling about the premises. Perhaps people kept out of the way because they knew Pottie was there. At times, during very long Ndumu sessions, when my bodyguard went up to Joburg for supplies, he always had Jack sleep on the veranda-dormitory with me. "He has killed before," said Pottie, "and he'd kill again to protect you."

The dawn cacophony of birds began at roughly the same time for all species, as if the first one awake and singing or squawking automatically roused all the others. At long last I must give the Purple-crested Turaco (or Loerie, as Afrikaners name it) due recognition on the Ndumu list. Though it was a moderately difficult fowl to see, its voice carried far, indicating a plentiful local population of its lovely kind. But what an utterance! Roberts puts it as "a series of loud, explosive 'krooks,' which precede shrill wailing notes." I quickly learned to pick out the voices of loeries, as well as those of doves and a few of the commoner barbets, shrikes and sunbirds. But it amazed me how many others Bruce could identify when he happened to be there. Sometimes I almost felt he must be inventing notes from thin air. But when he would mention a White-browed Robin, for example, and I would jump from my cot, exclaiming that that was a bird I hadn't seen, he would grab his shoes in the same instant and in our pajamas we would pick our way through the thornbush until he took me up to the bird.

One morning it was quite different. Pottie and I were enjoying our tea and talking about nothing in particular when four of the mosquito catchers' various younger sisters appeared alongside Jack's shack, about twenty-five yards away. These girls rarely appeared at the NRC camp and I had never seen them at this time of day. They seemed somewhat dressed up, wearing bright-colored blouses and with their hair done up in strings of tiny white beads. Whatever their purpose, they did a lot of giggling about it and often looked in our direction, for of course Pottie and I were completely visible in our cots on the screened sleeping porch.

"Come here," Pottie called to them in Zulu. "If you have something to say, tell me about it."

They walked over obediently, though still shyly and without decision. In front of the screen they dropped to their knees respectfully, in which posture the boldest and best-looking one—whom I now recognized as Dom-Dom's sister—spoke for them all.

The conversation took at least ten minutes, during which Pottie made several brief speeches and the girls nodded their heads in understanding. Then they arose, thanked him and walked away with perfect composure.

Not until the girls had disappeared did Pottie allow his features to relax into a smile. "They said they would be *honored* to go to bed with us," he reported. "Two for you and two for me. They think we are some kind of gods, and I had a hard time telling them that they did not understand the rest of the world, where there are many people like us, most of them unkind and unfriendly. I told them also that it would be against the rules of the game reserve for us to sleep with them. That was something they could appreciate. I told them we both were flattered by their kindness and were sorry not to accept their offer."

Indeed, the mosquito boys thought of us as deities, too. One morning Pottie returned from a catching session bursting with laughter. "I guess I was a bit noisy," he said. "It must have been all those beans. Anyhow, one of the smaller boys turned in wonder to Magwalo and said, 'I didn't know a white man could fart!'" Their sisters were more generous in crediting us with mortal functions.

I was working the mosquito boys harder than ever, and though I may have intimated that routine duties monopolized my time, I continued simultaneously to push the aquatic studies as vigorously as possible. To turn up odd mosquitoes that might give us new leads on viral bypasses, I had the boys concentrate on ground holes in the forest bordering Shokwe Pan and they did a fine job of capturing esoteric strays (devoid of viruses, however). In my own collecting I naturally scooped up mosquito larvae indiscriminately, along with other swimming creatures. When I reared these to adults in my insectary, I was astonished to find that in almost every case they yielded mosquitoes that differed from any we had caught on the wing. Thus mysteries connected with water seemed to increase in number and complexity.

And I had now become thoroughly lost among nymphs of mayflies, dragonflies and damselflies; larvae of various beetles, midges and other two-winged flies; water-inhabiting adults of water striders, back swimmers, creeping waterbugs, water scorpions, giant waterbugs, water boatmen, water scavenger beetles, predaceous diving beetles; and further aquatic fauna too obscure or too bizarre to bear common English names. I was not only out of my depth but also in over my head and almost out of my mind.

There were ever so many reasons not to give up, not even counting stubbornness. By definition the project was a long shot—years of effort might not be enough, and I had scarcely more than begun. But from the outset I kept getting encouragement from the lab, not the back-slapping or wordy kind, but the promise of early success. "One mouse group looks mighty suspicious this morning," Bob might write. Then, after I had waited with almost unbearable impatience for an entire week, next Saturday's airlift would bring the dry ice and another letter from Bob in which he might not even mention that "suspicious" mouse family. Later, when I would ask him about it, he would look into his records to search for what he had meant.

"It was probably a nonspecific reaction, not a virus," he would say. "Oh, yes, here it is. They were inoculated with a suspension from a giant waterbug, and all the babies died next day. We weren't able to demonstrate a virus by passage of brain material from dead babies to further mouse groups and Hugh suggested that the suspension must have been toxic. Do you agree?"

From what I had read of the bites of giant waterbugs, I most certainly did agree. Of course they have large sharp proboscides which can be mechanically painful enough in themselves, but those weapons are nevertheless backed up by sturdy salivary glands that secrete a copious and potent juice. If injection of such venom can bring curses of agony from strong men who step on giant waterbugs while wading, what could the substance do to the brain of a day-old mouselet whose skull hasn't even yet ossified? Toxic indeed! It was a wonder the babies had lived until the following afternoon instead of expiring before the needle was withdrawn. They are, for a fact, a lot tougher than they look.

It transpired that most of the creatures I sent to the lab were toxic in greater degree than mosquitoes, leading everyone to entertain a series of false hopes at the beginning of the venture. Even such bland invertebrates as snails evoked suspicious signs in infant mice, which means that babies displayed unnatural tremors, failed to nurse, or struck trained observers like Bob and Bruce as simply "not being quite right." To overcome those spurious reactions, Bob began to dilute the noxious suspensions, but of course that reduced concentrations of hypothetical viruses also, thereby weakening our chances of detecting them.

However, now and then a family of baby mice did strange things that did *not* seem attributable to poisoning, and though viruses were not recovered from them, Bob would feel that we ought to repeat the experiment. "Get some more of whatever it was you collected on such-and-such a date and labeled 27-D," he would write. My notes might show "27-D" to have been snails derived from a pond that was now dry, or else tadpoles from a ditch which now had been abandoned by baby frogs or toads. Perhaps "27-D" referred to water mites or dragonfly nymphs that *could* still be recovered from a shaded pool connected with Shokwe Pan, but there was no way of knowing whether they were the same kinds as originally collected. I had to face the very plausible reality that it was impossible for anyone to become an expert aquatic biologist overnight. We even sent for Dr. Arthur D. Harrison of the National Institute for Water Research in Pretoria, who came to Ndumu for several days to attempt that transformation for me, but I remained as before—at home in my tiny mosquito niche, otherwise utterly lost in the vastness of a liquid world which I had failed to tour when I should have despite the invitations of professors long ago.

People at the lab, by the way, were not getting along much more happily, as their experience with the more recent behavior of our golf-course viruses will illustrate. The first, Witwatersrand virus, is one of those curiosities of arbovirology—a virus which appears in the laboratory once as a mosquito isolate and then decides that it has revealed enough of its private life, locally that is. Where it had come from and where it would have gone but for our interfering mosquito catchers remains one of nature's secrets. Yet there must have been a discoverable story confronting us in the environment, for neutralization tests on 712 human sera from the nation revealed two positive reactors. Somewhere, even though rarely, man made the critical contact during which the virus passed from its wild harborage into his bloodstream. Later the ABVRU laboratory detected three reactors among 79 sera from Moçambique. Therefore Witwatersrand virus was not entirely a focal curiosity around the Germiston golf course.

And that brings up the real teaser. Germiston virus showed itself in two mosquito isolates before going into eclipse. A study of 367 human sera from the Highveld of South Africa, collected at that time, failed to reveal any reactors at all. But a small percentage of sheep, goats and cattle and twelve out of twenty-three equines did react. That would seem to point clearly to a veterinary problem. But elsewhere in South Africa, as well as in Angola and Bechuanaland, some 10 to 15 percent of the human population possessed Germiston virus-neutralizing antibodies. What can you make of such irregularity? You can't rightfully criticize nature for your own fragmentary information: all it means is that you don't know enough—indeed you probably know scarcely anything.

I really did not know where to stand now—whether Entebbe or Ndumu offered the better chance to stalk nature's secrets. The trouble was that one never saw a target, and stalking had to be random. No general would ever give so ridiculous an order as "Random march!" He would sit down and wait for something to happen that might influence his bump of decision. Perhaps I ought to do that, too. Instead of worrying about 27-D or 43-Q, maybe I should pay greater attention to birds. These were, after all, my consultants and augurs, and I had been neglecting them.

Of course I recognized this kind of thinking as the equivalent of thumb-sucking, but then, what dreams have not been initiated in such oral reveries, later to solidify in useful shapes? Therefore, let's dispense with excuses and be honest for once, admitting the following propositions as unalterable truth: (1) I would rather look at birds than anything else, and all my prattle about their giving me environmental orientation had been hogwash; (2) the aquatic viral hypothesis by this time was ready for the negative-results junk pile, not because it had been disproved but because the difficulties of demonstrating it had been underlined sufficiently to warrant its abandonment.

If I could really accept those naked statements, I should be ready to play ornithological hooky with the insouciance of Sandy Haddow. But I was not that honest, or else I was too conscientious. As I looked through binoculars guiltily, I vainly sought free-flying feathered darts that coursed through apparently frictionless atmosphere. Instead I saw them as though they swam laboriously through green, slimy muck. Rather than bearing feathers and singing with attractive beaks agape, they were clothed in snaky algal strands and they charged each other with jagged, rapacious jaws.

More illusions! Now I recognized the madness wish, for only if deranged could I be excused for having committed the cardinal scientific sin: *I had fallen in love with a hypothesis.* Subjective science is non-science. Emotion has no place in logic. I must exorcise the *desire* that my hypothesis be proved; if I could not thus repudiate all motivations for self-gratification, I should cut my intellectual throat. All right then: let's see what those neglected birds were doing, but let's look at them in the pure light of bird-watching, with no pretense of exploiting them for ecological witchcraft.

And what a relief that was. For there they were as I had first known them, sufficient unto themselves and unto me in their uncomplicated selves. The dry season had been exceptionally severe and prolonged, so that brush fires both within and surrounding the game reserve were a daily—and nightly—sight. Cattle Egrets and numerous large birds of prey, among which I recognized the rather nondescript Wahlberg's Eagle, converged on the smoking fronts chiefly, it seemed to me, to prey on insects that flew before the conflagrations, but they may as well have lit to feed on roast carcasses in their wakes. Fork-tailed Drongos, glossy black fly-catcherlike birds, also joined in the hawking. (I surprised the head ranger several times where I was sure fires had just begun, and he invariably imprecated a mercurial native whom I never saw, though I felt certain the ranger had struck the match.)

But this was heartily salubrious. If I stole as little as five minutes between mosquito batches, I could always run out and see something interesting near the camp. I would be especially happy if a male Plum-colored Starling had happened to alight on one of the trees in the compound, for these birds are clad, except for white breasts and bellies, with the most unbelievable hue of mulberry or, as described in one of my books, "metallic amethyst-purple, reflecting in some lights purplish-plum colour and golden violet." Or I might run into a small flock of White Helmet-shrikes, itinerant gray-black-and-white creatures with legs and circular eye wattles a bright orange. These shrikes seemed so happy in each other's company and in their foraging activities that they paid little notice to an observer and I could often approach them very closely. However, they always kept moving toward some goal, for despite their lack of apparent purpose in random insect catching, suddenly they would be gone.

Taking a few more moments of leisure than usual, I might stray beyond the NRC confines, occasionally frightening a basking monitor lizard near Mamba's marula tree and being myself thoroughly frightened by its thrashing exodus into the bush. Soon I would come to a fine fever tree growing solitarily in a gully that held water long enough in the rainy season to support this thirsty sipper. Here a male Diederick Cuckoo had established its calling perches. This tanager-sized bird belonged to one of the three resplendent species of *Chrysococcyx* to be found at Ndumu. With only minor differences among themselves, the males can be recognized by their brilliant metallic green upper plumage.

I became greatly excited one day when I saw a female Diederik appear in the fever tree, obviously in response to the male's self-advertising. Her arrival was, of course, to be expected, but the male's behavior then utterly electrified me. Within two minutes he fed her three caterpillars, bending sharply up and down before her once or twice each time. The plain-colored hen accepted the food readily and ate it, though she did not "beg" or exhibit wing quivering as young birds often do. This was obviously "courtship feeding," as ornithologists have named it, but what made it noteworthy was its practice by a parasitic species of bird. At least one school believes that feeding of the female during courtship, as a pure luxury like breakfast in bed, may have had its origin in the more utilitarian no-nonsense ritual of feeding her during incubation. In this parasitic species, whose females lay their eggs in other birds' nests, the behavior was purely atavistic. I thought this was a great discovery until I found that Roberts mentions it in his account of the Diederik Cuckoo, adding that "Adult cuckoos have been seen feeding young cuckoos reared by their hosts."

That was even odder! But so were countless other birds and their doings all around me. Near Site 18, for example, I used to see some small, ordinary-looking swallows that impressed me vaguely as ground feeders because they sometimes alighted on an open stretch of sandy soil supporting a few low weeds and bushes. Once or twice I went to the spot where I had seen a bird come down, to learn what had attracted it, but I could not find the swallow and assumed that it had taken flight without my seeing it. After this had happened several times, something in my mind signaled an alert and I began to concentrate on the problem. It did not take long to ascertain that I was dealing with a new kind of swallow (a life-lister), the Gray-rumped Swallow, which nests in old rodent burrows on flat ground. I had actually found the holes but quite understandably had paid no attention to them.

White-eared Barbets, Paradise Whydahs, Scimitarbill Wood-hoopoes—those could lead one in endless circles of local discovery. They did wonders for my mental health. Until . . .

I can't remember who found that little sheet and gave it to me. Probably it was Botha de Meillon. It was a one-page reprint of an article that appeared in the *East African Medical Journal* for January 1957. By H. Goiny, E. C. C. van Someren and R. B. Heisch, it was called "The Eggs of *Aedes* (*Skusea*) *pembaensis* Discovered on Crabs." My God! Here was the fabulous "nutshell" one had imagined: crabs, mosquito larvae and viruses all cozy together at the bottom of an aquatic crab hole. I must get back to Lumbo as quickly as possible.

21

PROFESSIONAL ORNITHOLOGISTS

Bob Kokernot supported the idea at once. At least he supported *me*, which was what mattered. Bruce McIntosh was conservatively skeptical as usual, while Paul Weinbren, seeing a lark in the offing, volunteered immediately to go along as official photographer. "I missed the other Lumbo expedition," he reminded us, "so I really ought to—"

"Now wait a minute, everybody," interrupted Bob. "This isn't going to be an expedition. There isn't enough time left, and the budget couldn't afford one anyhow. Whoever goes will do so more like an ordinary tourist than as a fully supported arm of this lab. Besides, as for time—Brooke ought to wait until the rainy season begins, as we did last year. We know that viruses were being carried about then and were probably being transmitted at that period. Well, if we choose late March again, we may have a chance of hitting some viruses in addition to making valuable comparisons with a known former episode of activity."

That delay took on a strange complexion when Bob added that meanwhile he wanted me to go to Kruger National Park, as if every time I had an exciting venture in mind I must sit down and watch hippopotami until I cooled off. I could readily understand his reasons for holding back on Lumbo for a while, but Kruger?

There was much too much going on at Ndumu for that—even Lumbo would be worked in only by sacrificing precious final days with my beloved *Aedes circumluteolus,* and I told Bob, without waiting to hear the rest of his proposition, that Kruger was definitely *out.*

"But I *order* you to go," insisted Bob. "Wil Downs is coming, and we've *got* to entertain him. He likes birds, but I don't know one from another, so it'll be up to you to show him around."

"What about Bruce?" I asked. "He likes birds, too, but he always seems to get left out of things. Let him take Wil around. And why must it be Kruger National Park? There are plenty of birds right around Joburg to occupy a visitor on his first trip to Africa."

Those questions led to a lot of explanations, to which I shall add a few additional ones. Dr. Wilbur G. Downs was one of our Rockefeller Foundation colleagues, currently serving as director of the Trinidad Regional Virus Laboratory in the West Indies. He had recently been put in charge of the Foundation's global arbovirus program and in that capacity was planning to make his first inspection of our South African setup. It so happened that Bob Kokernot's annual local leave was coming up at the exact time of Wil's suddenly announced visit. Bob was a great fan of Kruger National Park and had made reservations for Edith, the children and himself months in advance. He could not bring himself to give up the precious accommodations. Thus, when Wil arrived, Bob would come to Joburg long enough to pick up his guest and return with him to the park where they would hold their conferences; since Bob would be away from the lab, Bruce would have to remain as usual to cope with virus inoculations and "reading" daily mouse reactions; then who but CBW was left to point out dickybirds to the distinguished foreign chief?

Poor Bruce. He was doing his damnedest to be an ornithologist, and since that sounds an odd thing to say, I must give it some background. Finding himself approaching young middle age with an attractive family and a secure position as a veterinarian-scientist, he nevertheless felt not quite a "compleat" man because he lacked a hobby. He then put himself through a computer-type of analysis and came up with the conclusion that birds would be the proper avocation for a person of his age, position, disposition and so forth. Now, I have never before or since known of anyone who approached and then embraced the practice of ornithology by that avenue which, in truth, seems more like a rear alley than the front driveway. Other people elect birds without introspection or conscious effort—indeed they are usually helplessly impelled toward them, just as the bird-haters in our society are strongly but involuntarily motivated in their repudiation of feathery things.

Bruce had, in fact, come a long way in his study, not only by independent application to books and bird walks but also through membership in the Johannesburg Bird Club. However, he made no claim to being a professional. Attainment of such status, indeed, was not part of his plan, for that would have destroyed his amateur standing and then he would have had to look for a new hobby. Yet, despite his honest protestations, the ABVRU group had tapped him for the formal position of staff ornithologist, and willy-nilly he was now a professional in a strained sense of the word.

Obviously there ought to have been perquisites, privileges and other advantages to accompany such responsibilities, and getting away from the lab once in a while to do a bit of field work was surely one of them. But Bruce seemed to lose out every time, and he and I eventually worked up a perpetual joke or password between us to celebrate his chronic bad luck. We had heard of a Pacific Science Congress to be held in 1961 in Honolulu, at which time a panel convened from the entire world would discuss the intercontinental spread of arboviruses by migrating birds. It seemed to us that Bruce ought to be appointed to the panel to represent South Africa, though the chances of that happening were only imaginary. Therefore whenever we met on social occasions, we would have at least one secret toast to "Honolulu," were it no more than a matter of lip-reading across a room.

Bob was now forcing me into a professional ornithological role, too. Perhaps one can laugh at the idea that showing Wil Downs a few birds was "professional," but if this had been golf or bowling, I would surely have lost my amateur's classification, for the Rockefeller Foundation continued to pay me during "working" hours and I could not deny that for the first time in my life I was making a living out of bird watching.

We "did" almost the entire park, taking five days to comb it from north to south. I was amazed to see how great its regional differences were, for though two hundred miles is a goodly distance, it somehow still seems only a few steps when one is confronted by the vastness of Africa, and one would not be astonished to find the flora and fauna uniform within a compass as small as Kruger manages to embrace. Though I am not a botanist, it was easy for me to appreciate that we passed through a number of vegetational zones, in each of which both trees and more lowly shrubs and leafy plants displayed their special characters. Naturally the herbivorous animals distributed themselves according to their individual tastes, so that one found certain antelopes common in only one or two zones. Giraffes, which Merida and I had seen several times near Skukuza Camp, were absent in the northern end of the park, while elephants reversed that score. However, lions and impalas seemed to abound everywhere, and though I was no more successful in seeing lions this time than the last, at least I *heard* them roaring soon after our arrival in camp at Punda Maria. I noted that the gates had just been closed and could not help wondering whether the lions were pacing back and forth before them in disappointment.

The assignment was a cinch in several respects, though it had one formidable drawback as well. My guest, boss, colleague—or whatever he was from moment to moment—was totally unfamiliar with local South African birds, so that he needed an identification for each one in order to enter it on his life-list. Although he had traveled widely in other parts of the world, his enthusiasm on seeing novelties had not palled: the commonest species I could show him evoked an undiminished response, by which I mean that he exulted with the lavish happiness of youth rather than having that matter-of-course acceptance that may betray and befoul middle age. And Wil was interested in everything: mammals, trees, reptiles and flowers as well as birds; and photography, in addition to mere gazing.

What hampered me—as it had at Skukuza—was the prohibition on walking along the roads except within camp boundaries. Automobiles have not done as much to change the character of birdwatching during the past generation as you might suspect. They can be commended largely for conveying observers to widely separated lookouts during a single day, but most of the stops are not worthwhile if one is prevented from continuing on foot. Wil and I *did* get out once or twice when we simply *had* to get a better look at something or other. Would you believe it—we were more afraid of being caught and fined by park guards than caught and eaten by hyenas!

Wherever we stopped, we occupied cabins next to the Kokernots and spent the evenings with them. But during the day Bob was out in his own car with his family, while I drove Wil in ABVRU's station wagon. Thus we had all sorts of notes to compare over our preprandial brandy and water while Edith managed to look after the kids and get dinner for everybody in the traditional but miraculous female manner. The main bird I failed to show Wil (important simply because he had for some reason singled it out as a specially desirable one) was a Ground Hornbill. As often seems to happen, this fairly common species suddenly became scarce wherever Wil went but practically begged to be run over by the Kokernots' car. Bob's daughters, Jan and Peggy, had great sport each evening teasing Wil about the number of Ground Hornbills they had seen that day, and if Wil had cried for the girls' entertainment, it might not have been wholly an act.

He need not have been too distressed, however, because in spite of our physical restriction to the station wagon, we managed to work up a fairly respectable bird list—at least, a hundred and six species seemed a reasonable tally to me, while for Wil it must have been a bewildering if not actually staggering number. Moreover, the list contained eighteen life-listers for my own book, each one a bonus that had not been envisioned in General Kokernot's marching orders.

We saw some very fine and memorable species. Late on our first afternoon, just before we had to withdraw into the safety of Punda Maria's compound, Wil called to me to stop the car. "I see some kind of fairly large bird sitting on an exposed dead stub," he said. Of course I was not going fast, but I did not stop abruptly either, because any sudden motion, whether of acceleration or deceleration, may startle wild things and panic them into retreat. Thus, as I slowly came to a standstill, it happened that a full rising moon posted itself almost directly behind the bird. The circular field of my binoculars then revealed a perfect example of calendar art—a glimpse of nature so simplified and idealized that it could hardly be acknowledged as real. A clear, rapidly fading blue sky held the moon and a gorgeous swallow-tailed Lilac-breasted Roller perched on the weathered and twisted stub. Wil was as moved as I and quickly substituted camera for binoculars in order to record the sight on film as well as in memory. But the eye and mind are still occasionally superior to mechanical and chemical aids. Though Wil got a fine picture of the roller, he told me long afterwards that for some occult reason the moon refused to register on that negative.

One of the easiest birds to find was the Long-tailed Starling. Members of this family, which includes mynas, seem very often to be preadapted to man's civilized ways, for as soon as man appears in the wilderness with such unnatural fare as sandwich scraps, starlings of one sort or another emerge from the bush to undertake an easier life. It could not have been many years ago that Long-tailed Starlings were completely ignorant of refined garbage, but at Punda Maria they behaved as if their ancestors had evolved on it. If we must have starlings in the United States, it is a pity that someone did not introduce this species, for with its shiny blue plumage and finely barred tail it is much more attractive than its European cousin.

But easiest of all was the ostrich! Just imagine writing the word "ostrich" in your bird list and meaning it! When we first saw them —a cock and three hens, as I remember—they did not exactly register in my mind. That is, they looked so natural and tame that I had no impression that differed from my experience on seeing captive ones at the zoo. I almost thought, Someone must be keeping ostriches around here, when the realization finally arrived that these were wild. And what a significant life-lister from the taxonomic standpoint: *Struthio camelus* was a new species, genus, family and order for me (just as a penguin will be, if I ever get to see one)!

Even a new family is cause for delicious swooning of a sensitive ornithologist, but after those ostriches I let a Double-banded Sandgrouse slip past me simply as a species life-lister. Not until I was bringing my card file up to date in my rondavel at the Garden Inn did I suddenly respond to the name: sandgrouse, *sandgrouse* —why, that was the sister family that, along with pigeons and doves, makes up the great columbiform order of birds! But I could not swoon now, for the *Pterocles bicinctus* had long since flown.

I'd like to talk about many other noteworthy birds, but that would not be fair to the time we spent on such creatures as antelopes and elephants. It is criminal to dismiss the Kori Bustard by name only, to lump kingfishers together by saying that we saw five kinds, and to maintain total silence regarding scores of other signal species. Let me at least record that the breaches we committed when we left the station wagon at the risk of being expelled from the park were occasioned by glimpses, first, of a flock of Red-billed Helmet-shrikes, and second, of a Pearl-spotted Owlet. Otherwise we could not have been certain of them, and our lives, remaining forever uninformed and consequently unsettled, might as well have been delivered up to hyenas.

Actually the Skukuza hyena episode was not repeated on this

trip. Zebras, kudus, baboons, warthogs, hippos and vervet monkeys reappeared along with the impalas I have already mentioned, but besides those and elephants, Wil and I added nine names to my former score sheet. If some visitors to the park scoffed at bird-watching, there were others even more benighted who would not stop to look at four-footed furry creatures smaller than a deer. Again I consider them the losers for nurturing that circumscribed attitude. I, at least, was quite happy to see a yellow-footed squirrel for the first time and to give it as bold a position on the list as any of the antelopes. That may have been educated prejudice on my part, for the squirrel really didn't do anything more remarkable than run across the road. Come to think of it, however, did antelopes do more than that?

But when it came to dwarf mongooses, I believe even a confirmed antelope devotee would have agreed that the show they put on was peerless, fit for the eyes of any man, regardless of what broader interests he might have.

We saw them first as several dark objects moving along an uneven flanking ditch. Not sure whether they were birds or mammals, I stopped the car so that we could steady our binoculars for a look. Then it became clear at once that we had happened on a family of mongooses. My experiences with tame individuals of this tribe, both in India and not far away in Lumbo, had not made me forget that these animals are cousin to martens and minks, wary fur-bearers that can move like flashes and bring out the acme of cunning in trappers who succeed in catching them. Wil must have had the same thoughts, for we didn't even warn each other to be quiet—we simply froze in our places, except that as the mongooses came nearer I adjusted the focus of my binoculars and Wil his movie camera's.

This particular weasel- or otterlike form (not your lumbering skunk-polecat variety, but sinuous) is graceful in its snaky way, for despite the shortness of legs, the head, neck, trunk and tail are joined with such limberness that they appear to slide through space in the same follow-my-leader fashion as the successive parts of a serpent. Perhaps it is owing in part to the low-slung aspect of the animals that they appear so long-bodied and long-necked, but in any case they do not in the least look like dachshunds. There is nothing of heavy, pathologic achondroplasia in their habitus; rather, they look as if tailored and streamlined for the type of nose-to-the-ground sleuthing predicated in their master design.

But the dwarf mongooses were not shy at all. As they drew ever closer, they must have become aware of the station wagon—perhaps also of its occupants. However, that did not deter them from their course along the rough embankment of the ditch, and they soon went right on past us, so close that we could have touched them with a fishing pole. They were of a uniform dark-brown color, almost black in certain lights, their fur looking well groomed and sleek. As for their "dwarf" size, I did not automatically think of them as being small, though my books give sixteen inches as maximum nose-to-tail length for a dwarf mongoose compared with almost twenty-two inches for a banded one (such as our pet in Moçambique), and that difference should be great enough to be appreciable in the field.

I am still kicking myself for not having counted the members of that procession. What we first took to be a family of dwarf mongooses must have been a small community of them or else a family reunion that included distant relatives. As they ran along in single file, they appeared and disappeared along and behind hummocks of the ditch, so that one could not see all of them simultaneously. Indeed, they produced a total effect as of a huge, moving, banded boa or python, each mongoose being a black spot on the snake's back as its undulations flowed over the uneven ground. I now estimate that there may have been fifteen or sixteen animals in the group. Since most mongoose litters are small, averaging two or three kits, one can conclude that this species is socially inclined.

It is pleasant to believe that the leader was a wise animal, though of course it may have been only venturesome or domineering. For whatever it did was repeated faithfully by each mongoose in turn behind it, so that if it were a dunce, it was training its fellows to be dunces as well. I fail to see that this was a good method of hunting, so perhaps that is not what it was—possibly they were only going somewhere. If this were a prowl for food, the first animal would get the most and the choicest tidbits, and later ones would be better off to have hunted independently. If this were only a general search, to be climaxed by convergence of all the hunters on a termite mound, for example, the entire group would perform more efficiently by moving ahead on a broad front rather than one behind the other.

But it is rash to criticize or become skeptical on such short acquaintance. After all, we had only just met, and I must say in fairness that the last mongoose in line looked as active and well nourished as the first. After they had disappeared, Wil and I wondered all sorts of things about them, but most of all we bemoaned the sun's glare on the windshield: maybe the pictures would not come out. One question we settled quickly; the book stated that dwarf mongooses are chiefly diurnal and insectivorous, so that the animals had a right to be abroad in daylight and were probably hunting in a way that must be successful, no matter what our contrary opinions about it. Much later, the returned film of the sun's glare was adjudged splendid by viewers who wanted to know what the dark objects, vaguely seen behind it, might be.

Perhaps the antelope-fixation of many park visitors results from sublimation of the hunting instinct. That would account also for the general popularity of elephants and large beasts of prey. If so, then the mongoose would be a good test object for screening people into categories of pure huntsmen on the one hand and animal lovers in a more abstract sense on the other. Not that one accomplishes anything immediately useful in thus dividing the human race into factions, though it may indicate the degree to which we remain atavistic.

Wil and I saw some perfectly magnificent antelopes. The sable, the roan and the nyala can't be beat as trophies; besides which, they provide mounds of meat for the table. But I am afraid we did not have the correct attitude. While we were glad to see those proud representatives of their family, we were not made to drool by them and we gave equal thanks to the ridge-horned tsesseby and inconspicuous steenbok when they came our way. In fact we felt particularly self-satisfied after identifying the latter two, for these at least challenged us with the difficulty of determining their names, while more striking forms almost stepped out of their pictures on the page to greet us. And when we visited a lookout point along a river near Punda Maria to see the famous wallowing hippos and basking crocodiles, that did not prevent either of us from spotting and chalking up a Narina Trogon, gaudy in its burnished green breast and red belly, that sat quietly on a shaded branch overlooking the entire fauna, human beings included.

I remained a "professional" ornithologist to the best of my capabilities. Only when Wil took time off from birds himself did I feel free to indulge other interests. For example, several superb baobab trees not far from Punda Maria appeared before us on one of our sorties while we were talking about other things, and we stopped to have a good look at the monsters themselves, not caring for what they might or might not hold besides.

At one of the camps we got away from bird lore after Bob had introduced us to a German who was studying butterflies of the park. This amateur entomologist was experimenting with transparent plastic sheets of various kinds, trying to find the best kind between which the scales of butterflies' wings could be preserved. He carefully dismembered a butterfly, arranged the four wings in a natural position on one sheet, and then pressed a second sheet on top of them. On separation of the sheets, the scales were supposed to adhere, while the now naked membranous wings could be discarded. The sheets were then reunited and sealed. Now they could be bent, or packed tightly in layers, without harming the butterflies' impressions. It was certainly a much better way to assemble a collection than the old-fashioned method of accumulating bulky boxes containing horribly fragile pinned mounts. But it possessed disadvantages that even the enthusiastic German could see. For instance, none of the varieties of plastic he used was sufficiently clear. Furthermore, scales with pigmentary colors showed through well enough, but those whose metallic glints depended on diffraction of light lost most or all of their optical qualities. From a scientific standpoint, of course, the German's specimens were virtually worthless, since they were devoid of bodies and all-important appendages such as antennae and genitalia.

And how about aquatic viruses? If I had the boss all to myself all day long as we cruised about in the station wagon, wasn't this the time to clobber him mercilessly with my hypothesis? That might have been my course except for two reasons. Wil was scheduled to visit Ndumu later during his ABVRU visit, and I could more effectively give him my pitch on home ground, as it were. But apart from considerations of good sense and, perhaps, taste, I was enjoying my role as paid bird-walk leader too much to waste the hours in speculative dialogues.

A portion of the "professional" taint gained during that trip to Kruger National Park must have rubbed off, for in my next assignment, under no one else but Wil Downs, in Trinidad, I found myself bereft of entomology and encouraged to tackle small mammals and birds instead. As for Bruce, his brush with the profession must have been equally infectious. Some time during 1961, after I had become well oriented at the Trinidad Regional Virus Laboratory, I received a showy picture postcard from him. Displayed were two attractively garlanded brown-skinned maidens, while on the reverse side was the curt but exultant message, "Honolulu!"

22

HOTEL LUMBO

As Bob had said at once, this attempt to track a virus to an aquatic source in a crab hole was not meant to be a full-fledged expedition, and in many ways it was not, for no one was on hand to collect birds or rats or snakes or any other creatures apart from those that were my special interests: mosquitoes and crabs. Paul did let a few Vacutainer tubefuls of human blood, but that was merely by the bye: virologists let other people's blood freely wherever they are and whatever else they are doing. Paul's chief duties were to take pictures (as he had so certainly foreseen would be his happy function) and to help me with my peculiar needs.

However, I would be hard pressed to find a better word than "expedition" for the venture as it finally came into mature being. Based on the amount of necessary gear alone, that definition fit the situation, while correspondence, travel arrangements and movements of personel also attained volume or complexity, as the case might be, more befitting a serious scientific enterprise than mere tourism, as Bob had first proposed.

We initially became aware that this would be no simple outing when dry ice was mentioned. Of course, dry ice! Everything hinged on the preservation of viruses in whatever materials might

270

be collected. There was simply no way to make that part of it easy. You might cut down on everything else, but even if you reduced your objective to the virus-containing left hind leg of a single mosquito at Lumbo, you would still need the entire dry ice factory in Germiston and the internal airlines of South Africa and Moçambique to deliver that contaminated member in prime condition to ABVRU's laboratory workers in Joburg.

Once Bob realized how deeply we were committed on those scores, he more or less gave up on others, and from then on matters displayed a sort of runaway pattern in which one could no longer discern a formerly listed item named "budget." Well, that would be a pity for someone, no doubt, but it did not seem to bother anybody at ABVRU, and as far as I know our New York peers lost no sleep about it either. Perhaps the thing to do to avoid shackles is to find a good hot project and then ride it full steam ahead. If you look confident enough, people may give you credit for genius or some other valuable quality and overlook the fact that you spent a few thousand dollars to which you had no right.

Once again we sought and received help from the Instituto de Investigação Medica in Lourenço Marques. The arrangements were somewhat less elaborate than last year after all, for we did not need to ship anything by sea. Nor would we establish a camp at Lumbo: this time we would process materials at the hotel, provided that were permissible. Our chief need would be for a vehicle to transport mosquito catchers to and from various collecting sites, and to ply between the hotel and airstrip. If motorized transport of any sort whatever could be found at Lumbo, our "expedition" would be fully equipped. Jacinto de Sousa and an entomological technician, Simão Estephão, preceded me to Lumbo to see to those details as well as the hiring of field personnel. I, in turn, preceded Paul, for it should take him only a few days to obtain the movie sequences he wanted, while I felt I must have two solid weeks to collect a minimum mass of material for virus screening in inoculated mice.

You may consider hotel arrangements a dull topic, and so it is in most instances. Not in this case, however. João Branco, the manager, remembered us well from last year and his only regret was that the entire group had not returned. Business had been extremely slack. The railroad had in fact decided to close the hotel and transfer João very shortly to another branch. While he was not keen about that prospect, the move would at least rescue him from his present boredom. But right now, as of the current days and minutes, he was ready to turn over almost the entire hotel to us—not that it was at all vast, but in terms of public space he offered us the free convenience of nearly everything except a few square feet of standing area in front of the bar. We could even use his frozen-food locker to store unpleasant biological materials, provided we said nothing to anyone about it.

That turned out to be ideal. The lobby, dining room and bar were on an upper floor, beneath which at ground level was a large, screened areaway or veranda which we converted temporarily into a laboratory. Our intention had been to use our cramped bedrooms, or else to rent an extra room if one were available. But that would have put us out of sight and cheated João of all the fun and entertainment. As things worked out, he was able constantly to pass in and out through our working area on one pretext or another, until he finally gave up pretending to look for excuses and simply hung about as a frankly fascinated spectator.

A few other guests did come and go during the two weeks, so that João's time could not be entirely devoted to us. But he made no move to put us in the background when other persons had to pass through the areaway. If they were squeamish about mosquitoes or cyanide bottles or crabs, that was their hard luck. I think João would not have hesitated to hoist an ABVRU flag, if we had brought one, to advertise a not-to-be-missed attraction scheduled for a limited time only.

The luxury of an entomological technician was something not at all to my liking, for I was jealous of every mosquito identification, remembering Ken Smithburn's insistence on the professional seal to back up each one. But if I was going to study crabs, I obviously could not put in full days of mosquito work. Jacinto was an entomologist, too; his specialty was tsetse flies. And Simão had previously worked only in malaria programs, his expertise being consequently restricted to *Anopheles* species. The scheme we had worked out sounded fairly foolproof, however. Since *Aedes pembaensis* would probably again be by far the commonest mosquito at Lumbo, it being moreover rather outstandingly different from other culicines in that region, I would quickly train both Jacinto and Simão in the differentiation of this species from the others. That would leave me a greatly reduced load, once they had sorted and put aside odd specimens from the mass of *A. pembaensis*.

This year we were going to collect male mosquitoes, too, since the crab hypothesis involved infection of larvae regardless of sex and the subsequent emergence of virally contaminated males as well as females. Hence Jacinto and Simão must infallibly sort *A. pembaensis* into two categories for freezing in pure lots of one gender or the other, so that mouse inoculations at the lab could unquestionably be credited to notations on field labels.

I really hated to part with so much responsibility—not that I was infallible myself, but I had the feeling that because I cared so desperately about the results of this expedition, I would automatically be much more careful about avoiding mistakes than these two helpers who, conscientious as they undoubtedly were, had no realization of how critical each blessed mosquito might be in making history or perpetrating fraud. I was greatly pleased by the rapidity with which they caught on to my instruction, however. They mastered *Aedes pembaensis* males and females in short order and were soon asking to learn how to identify a dozen or so other species that were either fairly common or fairly unique in appearance. In the end I yielded almost the entire mosquito program to them, still feeling unhappy about doing so, but enabled thereby to further other explorations more widely than I would have thought possible.

It was good having Jacinto and Estephão working together, for they could check with each other when confusing specimens came along. Estephão, I should mention, was an *assimilado*, which means that he was a Portuguese African who had achieved a level of education sufficient for him to be assimilated into literate (meaning "white" without saying so) society. He could go anywhere and do anything, like any other first-class Portuguese citizen (though classes weren't officially recognized either). That he should have become an entomologist was somewhat unusual. But it made me think of poor little Qmba Ngwenya at Ndumu: that boy had entomological promise, too, but in South Africa it would be impossible for him to realize it (or for South Africa to profit by his talent).

Estephão spoke no English. Jacinto had not really come to identify mosquitoes but only to supervise the expedition in general, but since that included being interpreter for Estephão, he learned the work while translating my words. I once remarked to Jacinto that he spoke the best English of any Portuguese I had met and asked him where he had studied (knowing that he had not been abroad or attended a graduate school). His answer: "I spent all my money as a boy going to the movies in Lourenço Marques." What an ear he must have had!

Of course the first thing I had done on reaching Lumbo this time, even before unpacking and changing into field clothes, was to have the driver (whom Jacinto had procured by "stealing" him and his vehicle from a government station in nearby Nampula) take me to the edge of the lagoon where I might satisfy bursting curiosity with a quick look at a few crabs. The paper by Goiny and his associates had pictured eggs of *A. pembaensis* adhering to crabs' legs, on the conspicuous segment closest to the body. The species of crab most commonly infested, they said, had been identified for them at the British Museum (Natural History) in London as *Sesarma meinerti* de Man. That meant nothing to me, since I had never studied crabs, but the paper said nothing about *Sesarma meinerti*'s being anything like a hermit crab or a fiddler crab, and the illustrations of the legs, particularly of the front claw, did not suggest unusual anatomical configurations, so I assumed that I should look for the commonest, or at least the most ordinary-looking, crab to be found in the estuary.

Hermit crabs did not live in burrows anyhow. That left fiddlers, which were easily the most abundant but whose burrows were small and often very transient in appearance, and the large red-clawed crab that I remembered well from last year. This one might easily belong to the genus *Sesarma,* from the resemblance of its relatively undifferentiated claws to the pen-and-ink sketch reproduced in Goiny's paper, but of course it might be a different species from the one on the Kenya coast. Therefore—and nevertheless—that was the kind of crab I decided to catch and examine first.

What an admirable decision! But it was on my part only. The crabs had no thoughts of cooperating, so after chasing a few down their holes, I began to reconcile myself to the idea of waiting until next day, when my attire would allow digging. But that sounded like too easy a victory of crab over man, especially since this was the first skirmish and might set the tone for following ones. Bringing my dashes to a halt, I attempted to apply a superiorly organized nervous system to the problem and shortly devised a strategy by which I cut off a crab from its retreat with a bit of deft, rather than blundering, footwork. Now I was able to chase the crab toward high tide line, and as soon as it reached rough ground I stepped on it.

To my great satisfaction, I found this creature's legs lined with rows of small, dark, rod-shaped objects that looked exactly like the ones depicted in Goiny's article. Besides, they were most abundant on the front legs and on the segments next to the body, again tallying with the published report. I could not doubt that these were eggs of *A. pembaensis,* though I would still have to examine them under a dissecting microscope at the hotel before I could be certain.

That fact being confirmed not more than half an hour later—at least the rod-like objects resembled *empty* mosquito eggshells, and it was practically a certainty that there wouldn't be *two* exceptional species of mosquito, both laying eggs on crabs at Lumbo—the time for gloating had arrived. To have found eggs in abundance on the very first crab seized may have been a bit too generous a portion of beginner's luck, but it augured a successful expedition nonetheless, for if as few as 10 percent of crabs were infested, that would still bespeak an intimate association between crustaceans and mosquitoes, and the objectives of the expedition stood a chance of being accomplished.

One might wonder what was so great (in my opinion) about the eggs' being laid directly on crabs. We knew last year that *A. pembaensis* larvae developed in crab holes, so that they must at times come into contact with the holes' excavators. The larvae had to get down there somehow in the first place, and if female mosquitoes chose to let crabs transport them in embryo rather than descending egg-laden themselves into the holes to oviposit, what essential difference did that make?

Actually there was no concrete point I could make in answer to such a question. That was one of the troubles with the entire aquatic viral hypothesis—you could not drag it out onto solid ground. But I found a great wealth of *suggestion* in the *A. pembaensis-S. meinerti* situation. What I needed by way of anchorage was an indication that a mosquito and some kind of aquatic organism had formed an association—whether it were mutual or one-sided did not particularly matter—that had persisted for a long period of time. If a virus was ever going to establish itself as a Ping-Pong ball between two such hosts, it would most likely be able to do so if the hosts were in some fashion interdependent. A mosquito egg, deposited at large in Shokwe, would hatch into a larva that might swim the length of the pan (if it could) without coming into contact with some very specialized virus-harboring underwater organism. Evolutionary chapters might be written and erased under those circumstances without the development of a regular virus cycle between such isolated though contiguous hosts. Perhaps it was necessary to throw them together more intimately.

And, speaking of evolution, I regarded the laying of eggs on *Sesarma meinerti* as evidence of a very long association indeed. It could well be that *A. pembaensis* originally oviposited in small water collections of many sorts, crab holes being merely one kind that was utilized fortuitously. In fact, it would not have been necessary for females to go down to the end of deeper tunnels. Eggs deposited on moist sand at a hole's entrance would survive until they were knocked deeper by a crab's passage or perhaps were washed down by the incoming tide. But in those days (if this conjecture is correct) there was no certain prospect of profit when (or if) a crab-inhabiting virus transferred itself to a mosquito larva, for that host, upon becoming a winged adult, might never return to a crab hole, and most crabs might spend their lives without ever encountering mosquito larvae.

The crabs, as I learned during many hours of observation, came out of their holes in the daytime but did not wander far away, remaining ready to scuttle to safety at the first appearance of a large, moving form. At night they became bolder and foraged for greater distances. However, the state of the tide influenced their activity, too, so that occasionally one found them on the move in broad daylight. Mosquitoes seemed to time their activity along similar lines. *Aedes pembaensis* could be found biting in profusion along the lagoon on some days but only sparsely on others. Thus there were many occasions, between periods of their scavenging and flying, when the light was good enough for me to observe both crabs and mosquitoes sitting side by side in the entrances of the holes. It took a bit of slow motion to get that near without frightening the crabs, but I managed to do it several times, visualizing the mosquitoes at the closest focal setting of my binoculars.

Well, then, we have a nice neighborly scene, showing crabs and mosquitoes side by side on the muddy sand athwart a hole. The mosquito long ago used to lay its eggs on mud pellets but now it does so on crab claws. It seems a ridiculously short hop from the sand to the claw. How long did they thus sit an inch apart before the innovation was achieved? Although I can't answer for them, I believe they must have remained in their original positions, doing nothing, for absolute ages of time. Insects behave in response to instincts, and such patterns can never be changed rapidly (overnight is out of the question). Then, too, the first occasions when eggs were laid on crabs may have been unsuccessful in leading to the established custom. One can only surmise that this evolutionary change, or adaptation, or what you will, took as long as other manifestations of that plastic biological process. The average species, for instance, is thought to require about a million years to become distinguishable from its antecedent form. Perhaps *Aedes pembaensis* took a comparable length of time to fasten its reproductive destiny to *Sesarma meinerti*.

A million years? That, of course, was insignificant if it must all be occupied in taking just one step, for I wanted to proceed further and superimpose viral adaptations on the ovipositional one, and that might require a second million years! But there was no need to fret uselessly. As far as I knew, *A. pembaensis* may have been laying eggs on crabs for *ten* millions of years. Viruses then would have had all the time in the world to make what they would of it.

These thoughts were what had brought about our return to Lumbo. If ever there were a pinpointed situation in which my hypothesis might be demonstrated, the phoresy of mosquito eggs by a specific crab claimed that position. Phoresy again! I soon was enabled (in my most ragged trousers and sneakers) to make a census of that function by egg-bearing crabs.

The search disclosed that I had truly had beginner's luck when I caught that first crab. As I have previously described Lumbo's topography, the town sat on a low rise margined to the east by a bay of the Indian Ocean and on the west by an estuary that curved around the northern end of that spit of land, admitting tidal waters from the bay into the sheltered lagoon. Animals and plants adapted to the general environment attempted to colonize the entire littoral perimeter thus exposed, but the degree of their successes and failures was extraordinary in view of the slight physical differences that prevailed along the margin at its various fronts. At least the differences seemed minor to me, compared with the magnitude of responses to them. Wave action, direct exposure to wind and light intensity were the only three variable forces I could visualize, but they somehow conspired to prevent mangroves from establishing themselves where their maximum effects impinged. On the open bay side to the east, therefore, one found a few struggling small mangrove bushes that perennially failed to do more than spring up weakly and then collapse. As one skirted the northern point and turned westward, mangroves of greater stature marked the way until, at the most stagnant reaches, where the outgoing tide left flats clean-laved almost daily, mangroves spread into interdigitating thickets on luxurious tangles of stilt-roots.

If we now follow *Sesarma meinerti* along the same course, we find that a few of the crabs also colonized the bay shore of Lumbo. Here they appeared to be in some sort of difficulty, for they were not at all numerous and did not grow to full size. It looked to me as if they were unable to locate themselves in acceptable burrows—at least they had such holes scattered widely with respect both to high-tide line and the substrate. Apparently they liked to dig beneath and among solid objects (which I suppose made them less vulnerable to excavation by enemies). The roots of such small mangroves as grew there were scarcely adequate fortresses, and holes were more likely to be found under fallen palm trunks or occasional stones that were scattered on this side.

Again rounding the point, I noticed that as mangroves increased in size, crabs utilized them more and more routinely, until at the mangrove climax, *Sesarma* was wholly housed by the complex root system. A critical feature of mangroves, in *Sesarma's* ecology, is that they flourish best in the intertidal zone, which is exactly where this semiterrestrial crab must place its burrows.

One may suspect that another million-year courtship—or ten million—took place between mangroves and an ancestral marine crab that used to visit estuaries only at high tide but was compelled to vacate them when waters receded. It is easy to imagine the halting steps by which the cool burrow among roots, its depths still holding water, might have been discovered through a series of accidents and finally exploited *against* the better instincts of those long-extinct progenitors.

When I began to catch crabs right around the same circuit and examined their legs for the presence of mosquito eggs, I found it possible to make a perfect correlation between the abundance of eggs and the degree to which mangroves flourished at any given point along the littoral fringe. Thus, although the mosquitoes seemed adapted to crabs, they wanted their crabs *just so;* in other words, they preferred them to reside among thick mangroves in the estuary. One could almost be persuaded that the adaptation was really more to a particular sort of burrow than to the crab itself, except that then the mosquito might as well have omitted the crab in the first place rather than going to all the evolutionary trouble of becoming fixed to its portage.

Possibly the safest summary to make would include the dependence of A. *pembaensis* on *both* mangroves and crabs, each having its own share to contribute in a very special way, and A. *pembaensis* having nothing to do with either of them except when their favorable aspects happened to coincide. That was probably true on a broad geographic basis as well as locally in Lumbo, for A. *pembaensis* has been recorded as far south in Moçambique as mangroves and lagoons fringe the coast, but not beyond that point. I don't need to know what the southern limit of *Sesarma meinerti* may be: as long as the mangroves have petered out, that is sufficient to mark the end of the mosquito's range, but the same could work with equal validity if it were the crab that disappeared first.

Having learned this much about crabs—and also ascertaining that eggs could not be found on other kinds of decapods—I was now ready to trade on beginner's luck that had been transformed into knowledge of the whereabouts of the most heavily infested crabs. That margin of the lagoon—actually the favored latrine area of poor residents living close by—was home to some crabs that bore eggs on all pairs of legs as well as in crevices where legs joined the body and even on the eyestalks. Obviously this would be the best place to collect, from the standpoint of the frequency of mosquito-crab contacts. For there was another point to be considered. I have mentioned it in my ruminations at Ndumu but have omitted it thus far at Lumbo for the sake of simplicity. Possibly crabs were not sources of viruses at all, but they might nevertheless be agents in carrying eggs of *A. pembaensis* to a milieu where future larvae might pick up viruses from different sources. That might be from other burrow inhabitants, from animal food dragged into burrows by the crabs, from internal or external crab parasites, or from any of the manifold other contacts that crabs manage to effect during their daily and annual activities. Whatever the truth of it, whether viruses were acquired directly or by the most hidden labyrinthine means, egg-burdens of crabs could be treated as clues in the paper chase, and when I could divert my crew of boys to excavating burrows, I knew exactly where to tell them to dig.

By this time the first week had been expended in laying a solid comparative foundation of mosquito prevalence, so that I knew how each species was faring in March 1960, relative to its abundance a year ago. We had already shipped several thousand frozen *A. pembaensis* to the lab, where they had been inoculated into mice immediately in hope that I could be given early information while still at Lumbo, just in case that should be useful in directing further field operations. Paul Weinbren's arrival with his cameras and blood tubes coincided with the first report of "suspicious" mouse families. Good! That did not lead to any changes in field procedure, but it pepped up the program (if encouragement were needed) to know that viruses were again active. Our crab work would therefore be done against a background of high blood pressure.

We had sent frozen male mosquitoes, as well as females, to Jo-

hannesburg. Although the report from the lab was still only pre-
liminary, it did specify that none of the suspicious babies had received suspensions prepared from males. It would of course have
been exhilarating if the reverse were true. But since males had
been shipped in much smaller numbers, we could not afford to be
greatly depressed by the negative news. Remember that the hypothesis proposed that persistence of infection from the larval
state to adult mosquitoes of either sex must be rare, while adult
females could later acquire many additional infections by biting.
Thus, even if males had been inoculated into mice in equal numbers, we could reasonably have expected to draw a blank after
only one week of collecting.

Male mosquitoes are born into the world in numbers equal to
their sisters, but there are several physical reasons why they are
harder to collect. It is difficult to decide which factor is most important, for they all seem to work in equal degree against the man
with the net, aspirator device or simple tube trap. One obvious
limitation is simply that males do not live as long as females.
Thus, at times when the sexes are emerging from aquatic habitats
at a constant rate, females will soon outnumber males for the sole
reason that their population contains fair sprinklings of middle-
aged individuals and even oldsters, while the male contingent is
made up almost exclusively of young blades in their early prime.
This phenomenon seems to be related to diet. Males, devoid of
piercing mouthparts and equipped with oral appendages adapted
only to lapping, either don't get enough to eat from sweet plant
exudations, floral nectaries, rotting fruit, and the like, or else those
syrupy libations are not as sustaining as bellyfuls of blood taken
by their mates.

Then there are two related questions regarding accessibility of
males to the would-be collector. These again go back to the
matter of diet, in a way, because one reason female mosquitoes
seem easy to encounter is that they may be seeking the entomologist as eagerly as he is seeking them. While males may not go out
of their way to avoid a human being, they certainly have no reason

to move in his direction. The second part of this consideration is that males simply don't fly as far as females because they have no drive to do so. There will be plenty of vegetation near the breeding site from which they emerged, and there they are likely to remain. Unless the entomologist consciously specializes in collecting at the margins of centers of mosquito production, he will ordinarily find only a few straying males. Indeed the mating "swarms" which males of some species form to attract females are contingent on males' remaining together, rather than dispersing, and the handiest places for them to stage such gatherings are obviously alongside the home nursery.

My orders that males be collected on this second Lumbo expedition were therefore braver than they were sound. If we had not known the breeding site of *Aedes pembaensis,* the results would have been much more meager than they were. At the inland collecting site, one mile from the estuary, the boys found only one male for each 121 females collected! Thus, by the time they had accumulated slightly over 4,000 females (enough to inoculate almost a hundred mouse families), they had managed to catch only 27 miserable males (barely enough to inoculate one family). That was the total tally from the inland site, and you surely could not test a hypothesis concerning male mosquitoes on such a single-shot basis.

In the mangrove colony the situation was much more favorable. If we had collected there only from vegetation, i.e., foliage, twigs, trunks and exposed aerial roots of mangroves, we might still have been fairly short of males. But my experience in collecting mosquitoes from ground holes in the riverine forest bordering Shokwe Pan now gave me the idea of probing the crab burrows themselves. When crabs had darted into them, I sometimes thought I had seen vaguely moving shadows that could have been briefly disturbed mosquitoes. I soon discovered that if I poked a small stick into a burrow, jostling it roughly within the outer six inches or so, a few mosquitoes would fly out briefly. Some of these would alight on nearby objects before darting back into the holes, and during such brief halts they could be taken in tubes. The boys quickly caught on to that collecting technique, and I was happy to determine that a goodly fraction of mosquitoes inhabiting actual burrows were males. The ratio here was one male to slightly less than two females, which is about as high as you are likely to get anywhere, in view of the several forces we have reviewed that are going to keep females in excess under any conditions. Our final tally here was not greatly different from the inland site from the female standpoint, but we secured enough males to make tests in almost sixty mouse families. That might just about begin to be enough to mean something!

I had held off on crab collecting until Paul's arrival in accordance with a plan we had settled in Johannesburg. Obviously we had to get the mosquito background firmly in hand before crabs were disturbed: that I had accomplished during the first week. But crab excavating was going to raise hell with the appearance of the mangroves and the estuary, and Paul insisted that he must first get some movie shots of the scene before my invasion violated it. Paul's script or scenario or what-not went far beyond anything I had thought of in the beginning. He felt that we should have pictures not only of field-biological aspects that might have a relation to the aquatic hypothesis but a complete documentary on the life history of *Aedes* (*Skusea*) *pembaensis* (Theobald).

"We've got at least one new virus out of that species anyhow, whether your theory comes out or not," he observed. "From that standpoint alone it would be worth getting the whole story on film. But besides that, think of the opportunity *A. pembaensis* gives us, even if it had no medical importance at all. We've got everything in the photographer's favor—a spectacular life cycle involving crabs; a known breeding site, with a concentration of the mosquito population in accessible, flat terrain; actually a picturesque setting with excellent light (we can shoot from any angle with reference to the sun); and just the right humorous touch with all those human stools dispersed among the crab holes."

Regardless of how many people might or might not view stools as riotously funny, photography assumed a dominant place in the collecting program. There was a further reason to wait for Paul's advent, however, and that was because our crab collecting would have to be somewhat limited in any case. We would not have time to dig out innumerable burrows (comparable to our thousands of mosquitoes acquired), nor could the lab process an unlimited number of such relatively large organisms. For that matter, how *do* you grind up a four- or five-inch crab in order to inoculate it into mice? We had thought of this ahead of time and decided that the best thing to do would be to pool mere snippets from several crabs in one tube and submit this for processing. We might have one tube containing crabs' eyes and eyestalks, for example, another with pieces of gill, and perhaps a third with haemolymph (the "blood" of crabs) rather than solid tissues. Of course we would collect any other creatures that might be found in or near the burrows, so that further tubes might contain snails, marine worms or samples of cloudy tidewater that must be swarming with microscopic protozoa, algae, bacteria and so on.

Exclusive of the last-mentioned accessory samples, easily picked up in passing, we ultimately collected one hundred five *Sesarma* crabs, and *that* took some digging (and later some dissecting and pooling). Actually we approached another limit to our production schedule here, for by the last day we had nearly exhausted the supply of readily available crabs in the area of heavy egg infestation. Some of the burrows led under masses of mangrove roots that were so thick or tangled about stones (whose presence we had not suspected) that our boys were unable to reach their dead ends, and many a fat crab sat out the siege safely in its dark cul-de-sac.

The boys loved every part of it, even though they did not know what Paul was doing and probably thought he was mad. To begin with, he wanted a scene that showed them "going to work" (these and all subsequent instructions were relayed *via* Jacinto along with a continuous banter that needed no translation since it was so obviously part of his and everyone's good humor). They could

understand "going to work" all right, but then Paul—having taken
the shot as they arrived in the field van, climbed out, picked up
their buckets and shovels and plodded across the estuary to the
clumped mangroves—said he wanted to do the sequence again
from a different distance or perspective. Hence the boys must
come back, lay down their equipment, climb into the van, and
then "go to work" a second time just as they already had. Perhaps they even had to do it three times. Well, that was perfectly
acceptable, for it was easier than the actual work of digging and,
after Jacinto assured one of their anxious spokesmen that they
were being paid for this nonsense, they were more than happy
to comply with the most extreme aberrations that Paul could
dream up.

So it went with the rest of the operation—digging, chopping
through mangrove roots, digging further, and at last coming up
triumphantly with a crab. But what insane joy it was then to throw
the crab back into the hole and duplicate the thrill of catching it
without having to repeat all the effort! Those boys took to acting
as if it were in their blood, and who can say that it was not?

Back at the hotel we went through many additional histrionic
sessions, with mosquitoes as well as crabs putting on endless repeat performances at Paul's command. Crabs, now scrubbed clean
of their mud, were set down on an open, freshly swept walk by
the front steps, where they might be seen to best advantage. We
wanted to show several things: general shape and size, method of
locomotion, and so on, but especially the handsome red claws
with their rows of mosquito eggs close to the body. Paul had
equipped himself with all sorts of lenses, so that he was able to get
shots even closer than the human eye could focus. Of course, the
nearer he tried to take pictures, the more greatly crabs' movements were magnified, and since their idea in the first place was to
gratify our wishes for action photos by disappearing as fast as
they could, we were at length forced to take several crabs apart to
show some of the finer details of their structure and especially
their phoretic services for eggs of *A. pembaensis*.

Mosquitoes turned out to be traditionally recalcitrant and temperamental as actors, too. Paul wanted a sequence of the biting process, from insertion of a proboscis to the final state of complete engorgement. The insects seemed eager enough to bite while we were working on crabs among the mangroves, so they ought to bring their appetites with them to the hotel. We would unplug a cotton-stoppered tube, upend it so that the contained mosquito came to rest on the back of a hand, wait for the creature to begin probing, then quickly pull the tube away so that Paul could get a clear shot, undistorted by curving glass.

It soon appeared that the mosquitoes were not really as erratic as we thought at first. After many trials, a few of which resulted in biting but most of them in refusal to feed, we realized that practically all the failures took place when we were using Jacinto as the victim. He then told us that in general mosquitoes seldom bothered him. They did not like me very much either, though several less choosy ones were finally induced to partake of my substance. Moreover, neither Jacinto nor I felt anything while being bitten by this species, and no swelling or itching took place during or after the meal.

But Paul was apparently an entirely different piece of meat. *Aedes pembaensis* females could not get their beaks into him fast enough and he suffered not only while they did it but also for ten days afterwards. Allowing himself to be bitten was extremely inconvenient for movie making because he found it awkward to manipulate the camera with his right hand while focusing it on a mosquito biting his left, and we finally compromised on me as the subject to use for those shots.

I had no idea that this particular film strip would be so good. Indeed, I would have bet against it, because Paul and all his gadgets seemed to be almost on top of the mosquito, and I thought that the image must surely be faint and hazy. Not at all. He actually obtained several excellent sequences, and it was difficult to decide which one was best. The reasons for taking more than one included, naturally, the photographer's chronic fear that the first one might not come out properly. Multiple shots are the photographer's insurance. But in addition Paul wanted to end up the drama with a proper splash or climax, namely with a slap in which the engorged mosquito was exploded so that the parting scene would be that of a messy red splotch on the back of my hand. The trouble was that I waited too long to slap. The mosquito always caught me off guard, suddenly withdrawing its proboscis while I thought it was still going to take a few more swallows. Off it would fly with its belly full of free wine, and though its progress was now burdensomely laborious and I could easily have squashed it at its first landing place, that is not what Paul wanted: the end must be in the same pub as the beginning.

On the screen, *A. pembaensis* now showed her form and colors more strikingly than I had ever seen them through a hand lens. Paul could blow up the image until it was a yard across without losing sharpness of focus. We could see slight breezes nudge her antennae and daintily tip up her wing margins. Her sensitive proboscis tested my skin in several places until a suitable one was found. Then the piercing stylets were inserted in a series of thrusts, the encasing labium buckling back as the blades became progressively unsheathed and lost in a vast underground well of blood. After full extension of the mouth parts, the mosquito seemed to be doing nothing for several moments, but then at last her abdomen began to swell. Within a short time it looked fully distended, as if it must burst. However, the insect continued to drink, occasionally withdrawing her proboscis partly and then sinking it back to the hilt, as if to savor full satisfaction of the draught. Whatever had already occupied her digestive tract was now expelled as a few clear droplets, for no available space must go to waste—each extra swallow of blood will mean that many more future eggs. Finally a limit was reached: it *had* to be so, but how the mosquito gauged it so neatly, just this side of spontaneous rupture (when I needn't have slapped) was amazing to see on the screen, for now it looked as if that abdomen had expanded into a dappled goatskin containing several gallons of liquid.

Extraordinary as Paul's pictures were, we ended up with many very important gaps in the portrayal of A. *pembaensis*'s complete life history. We had no shots of mosquitoes' mating (which they may have done within crab burrows). Eggs on crabs' legs, yes— but the film did not show females depositing them (whereas Goiny and his colleagues had seen that act performed by captive specimens). The greatest omission in our series was in the larval and pupal stages. And not only on film—we had not found or collected a single aquatic form in the field. By the time burrows had been excavated in quest of crabs, the biological environment was so messed up that I presume larvae and pupae had been hopelessly destroyed.

On the other hand, there were some things that I would have considered extraneous or needlessly extravagant but which Paul insisted on including in his documentary classic. Lungfish, for example, had no claims on footage that I could see except that they were neighbors in some parts of *Sesarma*'s range, though that was out of the breeding milieu of A. *pembaensis*. But lungfish were both photogenic in their ugliness and romantic in their biological or evolutionary connotations, so Paul spent hours standing, knee-deep in water, filming fish as they climbed out of muddy tidal holes to sun themselves on old pilings and treat their primitive respiratory systems to clean, fresh air.

Then there was a shot of A. *pembaensis* that, while extremely memorable to us, was so personal as to have very limited appeal to anyone else, and surely it had no scientific value whatsoever. João Branco, more an enthusiast about our work with each passing day, decided to give us a great parting feast. For this he typed out an elaborate menu, each item being designated *à la* this and *au* that, such as salad dressing *au Weinbren* or *café à la Sesarma*. The entree was, of course, a huge rock lobster, similar to the one he had sent to our camp in the *chefe de pôsto*'s compound last year. Now, in advance of serving it, he had dubbed it *Lobster à la pembaensis,* and in order to make it authentic, he requested me to set a pinned mosquito in each eye. Paul's sequence here first shows João, dressed in chef's apron and hat, holding the lobster on a huge platter; then a close-up of the lobster after João had placed it on the table; and finally a further blow-up with a zoom lens: there, transfixed and transfigured, perched the essence of our expedition.

Naturally the film had its virological aspects, too—the sorting of mosquitoes, but, after that, their cyaniding, entubing, labeling and freezing in dry ice. The photogenic part of this sequence must be the departure of a little two-engined plane as the steel thermos flasks began their trek to the lab in Johannesburg. Paul got it all in, and the film could well be about any arbovirus field enterprise except for the parts featuring crabs.

One would not be far from the mark in observing that the expedition had been absolutely perfect from Paul's point of view. We both had to wait, of course, prior to passing respective judgements. His films were returned before the last of my mouse groups turned out negative, so he had already savored triumph when total defeat had to be acknowledged as my portion. The lab profited, in that six virus isolations were effected (four of them the same as the mysterious new virus encountered a year ago—the one that seemed to come from nowhere inasmuch as the mosquitoes had fed only on man but human beings showed no antibodies against the virus). But the six current isolations were *all from females*. Males, crab components (eyes, eyestalks, gills, haemolymph) and crab associates (two kinds of snail, two other kinds of crab, water samples from burrows)—every one of those contributions, even following a number of heart-stopping, red-herring, "suspicious" reactions, came up a good thumping solid negative.

Even last year's mystery was cleared up, if not beyond doubt, at least in highly suggestive form. Paul let blood from thirty residents of the Lumbo area, and this time four of them reacted positively to the virus, a significantly higher rate than last year. Moreover, of the six virus isolations effected on our second expedition, five came from female mosquitoes obtained at the collecting site near a human settlement one mile inland and only one from females taken in the mangrove breeding colony. This would make it appear that viruses had been acquired on the mainland, whether from human beings or not, and the mosquito yielding virus in the mangroves must have been a blood-engorged one that had returned from her own feeding expedition to oviposit in a natal crab burrow.

What could have gone wrong, then? We certainly had collected the right kind of crab. A pickled specimen was kindly identified for us as *Sesarma meinerti* de Man by Mr. V. G. James at the University of the Witwatersrand in Johannesburg. But that scarcely mattered, since we already knew that *A. pembaensis* used it for egg laying: whatever its species, it functioned in the intimate relationship I had predicated as necessary for a mosquito larval associate in an aquatic arbovirus cycle. The only remaining explanation for failure had to do with numbers. But how many male mosquitoes and what quantities of crabs must we process before nature would yield up the one in a thousand or one in a million that knocked down a mouselet beyond any question of trickery? If it were all a matter of odds and the long shot, why couldn't we have had a bit of early luck? For now the party was over and we could not stage a third lobster banquet at Hotel Lumbo.

That was all well enough said and properly lamented except for one thing. If the hypothesis were invalid, we could process male mosquitoes and dismembered crabs for the rest of our lives, to confirm in perpetuity what had already been demonstrated on a small scale: viruses do *not* frequent those hosts. Obviously not even a fanatic would go to such lengths—he would have blown his brains out after the first ton of mosquitoes and freight carload of crabs. But even the choice of that exact moment of despair would have given him trouble, for in biological research it is sometimes practically impossible to tell when you have adequately pinned down a negative quantity. I envy the fortunate chemist, physicist, engineer, mathematician or other dealer in inert forces and concepts who can see minus and plus with equal clarity. The student of life is confronted with a panorama of positives, so teeming with factual items that one wonders why he should bother about what isn't there. In a sense, he doesn't do exactly that, but in trying to explain visible things, he must postulate many invisible ones, and his imagination may go awry. . . . What, by the way, does a glowworm see with its taillight?

23

THE HOUSEFLIES OF ANGOLA

Bob Kokernot had said there was scarcely enough money for my expedition to Lumbo, and here, *following* that expensive venture, he was briefing me as his traveling companion on an aerial jaunt over the entire province of Angola.

"Oh," he explained when I pointed out what I thought was an inconsistency in his bookkeeping, "I had already planned this expedition. In fact that is why the budget was so low when we were talking about Lumbo, because for these four weeks just ahead we're going to be flying all over the place, staying in hotels on many nights, and that costs *real* money."

His purpose on this jamboree was to make a last-ditch collection of human serum specimens to leave as a sort of legacy to ABVRU, so that after we had both departed, the remaining staff could think of us as they performed antibody neutralization tests with the sera in baby mice, I suppose. The reasons for my going along might have been successfully challenged, had anyone been interested in doing so, for Angola in June would be experiencing its winter and mosquitoes would be scarce. Besides, if I had only a few hours at each stop while Bob bled a couple of dozen people, I could obtain only a minimum of specimens, if any, and form only the sketchiest of opinions as to the probable summer fauna of virus-transmitting insects.

Yet that is how surveys must sometimes be made when time is at a premium. Granted more days or weeks, they stop being surveys and take the form of legitimate studies. By implication, then, surveys are rather unsatisfactory at their briefest. However, one can justify them on the basis that they are better than nothing, and if Bob and I did not go now, during our next-to-last month in Africa, ABVRU would be deprived of the opportunity to estimate arbovirus activity in Angola, and that province might then remain unstudied from such a standpoint for another decade or more before any other laboratory interested itself in the situation.

Bob would not have mentioned it to anyone eying our budget, but he wanted me on the trip not so much as an entomologist as someone to talk to. Anybody would have satisfied him, as long as he could communicate, for that is how Bob put his thoughts in order. However, to state that he needed an entomologist sounded highly reasonable, and that is how I was chosen.

I was not at all averse to going, for at last I would have no scientific worries on the job. The aquatic hypothesis was done for. In Angola we would not collect mosquitoes for inoculation but only for identification. If I knew what they were on the spot, I would *throw them away!* If not, or if they were of interest simply as establishing new locality records, I would pin a few representative samples and bring them back to Joburg for further study by Jim Muspratt and eventual incorporation in the collection of the South African Institute for Medical Research.

So, in short, I was going along mostly for the fun of it, and that was that. Oh, I had considered all my African excursions as the greatest fun in the world, but those trips to Ndumu or Entebbe, Lumbo or Port Elizabeth were always dressed up in serious raiment behind which I could hide guilty personal exultation in being one of the party (or *the* party in the case of Entebbe). Now, with only the flimsiest of excuses to join Bob in rattling around Angola, I stood exposed as a scantily clad pretender, but *shameless* at last!

Hugh Paterson, in truth, almost lent me a light coat of legitimacy by urging that I collect houseflies wherever we went. "As far as I know," he said, "no one has ever collected them anywhere in Angola. Of course the province must be full of houseflies—they must be everywhere, just as they are in other places. But there is always the question of species in each locality."

Hugh suspected there might be some variation between the insect inhabitants of Angola's tropical lowlands and temperate highlands. Possibly *Musca sorbens* occupied the former and the more familiar *M. domestica* the latter. This was of some public-health significance, because the two species bred in different kinds of filth and therefore might carry either intestinal diseases or eye infections, depending on what sort of contamination dirtied their feet.

"Therefore any fly you can catch and label anywhere at all will be worthwhile," Hugh continued. "Just carry a few cotton-plugged tubes in your pocket constantly, and at the end of each day kill the flies in cyanide and mount them on pins with notes as to date and place of each capture."

Locality records—my list, as eventually given over to Hugh, sounded like the Portuguese version of some stationmaster announcing departure of a far-flung Paoli Local: "Béu, Buco Zau, Cabinda, Catete, Caxito, Cela, Dundo (or Portugalia), Lobito, Luanda, Luso, Malange, Maquela do Zombo, Moçâmedes, Nova Lisboa, Nova Redondo, Porto Amboim, Sá da Bandeira, São Salvador, Sazire, Silva Porto, Toto, Ucua and Vila Pereira de Eça! All aboard!"

Actually, although I knew nothing about houseflies (and really didn't want to get interested in them because I already had more hobbies than I could possibly handle), I was strongly motivated to do a good job for Hugh. I had failed him deplorably as a student of cow-pats in Uganda. But possibly I could make good on this final request and with these last four weeks in the field.

With an eye always on birds, with new people and places to meet and see almost every day, with such mosquitoes to identify as could be collected by hastily recruited man- or boy-power, and now with houseflies to procure in hotels, restaurants, clinics, airports and latrines, I found myself enormously entertained. Somehow I always manage to let someone else attend to the details of getting from one place to another. I don't mean to shirk; perhaps I simply look helpless. But Bob handled the money, tickets and human relations, while I seemed to drift along and respond to whatever cue presented itself wherever I miraculously happened to have landed: bird, food, mosquito, bed, fly, or that miserable Portuguese Constantino brandy. Of course Bob had corresponded about this expedition months before we began our trip, so much of its smoothness had been contrived. But it still required the good graces and exceptional kindness of Dr. Vitor M. R. Casaca, Acting Director of the Instituto de Investigação Medica de Angola, to make it as satisfactory as it turned out to be. In the first place, he put us up at the Instituto's offices while we were in Luanda, the provincial capital (saving oodles of money for the budget), and in the second he accompanied us on some of our excursions. We weren't able to cover Angola in a single circuit but had to keep coming back to Luanda to take off in each new direction. Thus we came to call the Instituto's headquarters our "Hotel de Investigaçãos."

The Portuguese seemed to regard French, rather than English, as their first foreign language. Vitor Casaca's presence for so much of the time was consequently an outstanding help, for many of our more remote contacts knew no English whatsoever. Even Vitor was not too good at it, and he quite charmingly kept introducing Bob with a French accent as "Doctor Kokernō." But from time to time we found ourselves on our own, and then when something went wrong with the arrangements we were really stuck. Of course in an impoverished land such as Angola, with a small struggling internal airline, one cannot expect anything but irregular service at the best. At one point, somewhat past the halfway mark of our trip and far down the register of Bob's good spirits, he told me that he had kept track of how many hours we had thus far spent in airports waiting for planes. The score then stood at thirty-eight hours. I forgot to ask him the eventual total.

One observation that impressed me more than ever was the irregularity of distribution of various kinds of organism—animal or plant—from place to place. That, incidentally, is an excellent argument for multiplying your interests as far as possible, and perhaps I should not have been as wary of embracing houseflies as I tried to be. At any rate, we did not strike a single place that was barren in all respects at once. If there were no mosquitoes, there would be birds or flies, or other features such as unfamiliar native trees and plants, agricultural practices and so on. And of course some places had everything, in which case we might be regretfully overwhelmed. The extreme possibilities were well represented on a wall painting of Angola province in the airport at Luanda. Lowland gorillas had been selected for depiction in the lush northern enclave of Cabinda; diamond mines along the northeastern boundary held in common with the Belgian Congo; but in desert country to the south, adjoining South-West Africa, the artist could find nothing better than that botanical phenomenon *Welwitschia mirabilis*—a recumbent plant of primitive affinities, bearing only two narrow leaves, the frayed tips of which, when mature, may extend ten feet apart from each other on the sand. Yet the leap from jungle apes to desert monstrosities was not altogether fantastic: Angola, no matter how minor its place on the map of Africa, is almost twice the size of Texas, and, as Bob would assure you, that's big enough to contain almost *anything*.

We saw none of such wonders. Indeed, I fear that poor Bob was so occupied with officials and with his bloodletting that he could rarely enjoy the scenery except from airplanes and occasionally from motor vehicles. Descending to the Luanda airstrip from the east, he must have observed the hundreds of young baobab trees growing in nearby wasteland. These seemed seldom to reach maturity, but as saplings they already gave promise of misshapen adulthood (except that in the case of baobabs this is *not* the wrong shape—it is the trees we call "normal" that are bizarre in *their* nation). The landscape, crowded with those green spears, looked foreign enough to belong to a different age. Bob enjoyed also the spectacular flight to Sá da Bandeira, which lies almost six thousand feet in the sky at the verge of an escarpment that dives down to the desert coastal plain. In Cabinda he may have found the tropical rain forest on the road to Buco Zau as stimulating as I did, but possibly for different reasons. While I reveled in reflecting that I was in the habitat of elephants, gorillas and chimpanzees, actually breathing the same air, Bob was worrying about what might have happened to blood tubes in the thermos flask when the Land Rover's steering mechanism came apart on the mountainside and we almost lost our lives as well as the specimens.

There were, in short, innumerable blockages against conscientious work with houseflies, though I deny that their optional ranking had anything to do with my neglect. I must come back to the concept of a "survey" to find a fair place in which to lodge the blame. When you are on a real expedition, such as our Lumbo enterprise, the novelties at the door of your camp (or hotel) wear off enough after a day or two to permit the eye to focus and the mind to concentrate on things within arm's reach. Otherwise the feet remain overstimulated and you have to keep your sights raised to avoid stumbling.

Not that I didn't catch flies: I caught *lots* of them. But I had to do it in passing, and I fear I succeeded in my design to learn nothing about them in the process. However, everything else also came and went as vignettes, as I have already suggested. I retain confused recollections of the margins of many towns and villages where mosquitoes were or were not taken; and, if they were present, the mosquitoes were either recognizable or not, or abundant or not. Very unsatisfactory. Among those glimpses I see occasional fertile irrigated oil-palm plantations with, appropriately enough, a Palm-nut Vulture appearing here and there. The first of that species I saw was a life-lister. As vultures go, it is a beauty of the tribe with its black-and-white pattern (including a banded tail), and it commends itself to our mores also by feeding on the outer pulp of oil-palm fruits rather than carrion. At least its diet is relatively "clean," where clean feeding is possible: I saw Palm-nut Vultures in many places remote from palm plantations, even at Buco Zau, high in the coastal rain forest, where I fear refuse provided by man was more likely its meat.

Buco Zau! The sound of that name reminds me of numerous disconnected figures which gradually coalesce to form a more substantial, composite vignette. We had slept there—somewhere —and had gotten breakfast—somehow. Now it was still early, with a heavy, cold mist shrouding the rim of tall jungle trees that encircled the small village. According to the schedule Bob had worked out with Vitor, natives from the general region had gathered, and now Bob screened them, with assistance from the local doctor, to determine which of the assemblage would be chosen for bleeding. I took a dozen ragged boys to a cocoa plantation along a nearby river and searched with them unsuccessfully for mosquitoes. The mist condensed on leaves, dripping on us in huge chilly blobs, and I was soon shivering within less than five degrees of the equator. A species of *Culicoides* (midges that are sometimes called sand flies) bit ferociously. Almost invisible, the tiny insects left perfectly circular red marks the size of a dime. Bob, with his blond skin, became wondrously polka-dotted. When the fog lifted, very suddenly, *Culicoides* all disappeared equally fast. A small flock of black-and-white swifts flew out of the jungle— another of the many life-listers I would not identify this side of heaven, I thought. However, Bruce had a book on Central African birds, and back in Joburg I ascertained that I had seen Sabine's Spinetails, belonging to the same genus as the Chimney Swift: I had been almost back to Pennsylvania in my gorilla habitat!

The best locality of all for flies was Toto, a spot not on our list for blood or mosquitoes but, according to our hosts at Maquela (where Bob *had* punctured a fragment of the population), "a good place to eat." That was indeed a widespread consensus, for the small, enormously hot, incredibly filthy restaurant was congested with humanity. Here flies were so abundant that it was almost impossible *not* to collect them. That they were all over the food goes without saying. But they were everywhere else as well —on tables, chairs, floors, walls, windows *and* people.

Whether those flies took part in attracting birds to the premises would be difficult for me to state with authority, but under the stone steps of an adjacent building was a colony of at least five hundred mud nests of Cliff Swallows. The site had been chosen for some good reason, whether I could name it or not. In their constant darting before the restaurant, the parent birds must have gleaned at least some flies from the air for their fledglings, though it is possible that an avian colony of that size may have provided flies with more than enough excrement for feeding to compensate for any moderate numbers consumed. My bet is that birds were secondary and that the restaurant deserved chief credit for engendering and sustaining those insects. I'll admit that I did not inspect the kitchen or check on garbage-disposal practices, but I'll take side bets about those, too.

My most difficult chase was in a hotel room at Nova Lisboa. Here the goddamn flies refused to settle on solid objects but hovered absolutely motionless at the center of things, just about over the middle of the bed. I had not thought to bring an insect net, so all I could do was try to hand-catch them or to swat them with a newspaper folded into a baton. The flies dodged with no seeming effort and immediately took up the positions from which I had ever so momentarily startled them. I spent energy, sweat, strength, profanity, patience and time on those flies, triumphantly catching one at last. I was advised later by Hugh Paterson that this was not a housefly and he was not interested in it.

On June 24, only six days before the Belgian Congo attained independence, Bob and I were in Léopoldville, having finished our work in Angola and stopping here for a day to make airplane connections on our way back to Johannesburg. The city was jammed with Belgians trying to get reservations to their homeland and with tribal chiefs from the wilds who had come in to see for the first time what it was they were about to inherit. The atmosphere was eerie and ominous. Still, there was no reason not to catch flies. "Hugh didn't mention the Congo," I told Bob, "but I might just as well pick up anything I can."

We went for dinner to a quite fashionable restaurant on an elegant street. Our table was not far from a tall window that gave out upon the sidewalk almost like a French door—at least it was a full-length piece of glass. With an empty tube in one hand and a cotton plug ready in the other, I chased flies on that window, bending up and down, up and down, as they ran and flew in short hops along the glass. Up and down, up and down. . . . Gradually I became aware of a cluster of African faces outside, watching me. They were solemn, wondering faces, perhaps even hostile, certainly not friendly. They belonged to people who had come to learn what sort of person the white man—the former master, the present fugitive—might be, what things he did. Suddenly I lost interest in flies.

"Come on, Bob," I said. "Let's get out of here."

24

I HAD BABIES,
THEY WERE STOLEN . . .

It was a pity to be leaving this wonderful project in this spectacular country. As a matter of fact I had been given the chance to prolong my assignment in South Africa for another two years. However, there was an attractive alternative—to join Wil Downs at one of the Foundation's other arbovirus projects on the tropical Caribbean island of Trinidad—while here in South Africa one had constantly to endure the poisoned atmosphere of apartheid. So I decided to go.

Of course I had missed a couple of long shots in my quest for aquatic sources of arboviruses, and that might have made Trinidad look greener after my failures at Ndumu and Lumbo. Actually I think my decision was not influenced by that. Looking at my work as objectively as possible, I could see that it would be better to stay in South Africa, where I could build on what had already happened, than to begin from scratch in a new environment. Despite the accomplishment of mistakes, I might already be closer to what I wanted to know in the Ndumu Game Reserve than if I walked as a stranger into some unfamiliar tropical American forest.

For we now had a splendid mass of positive information from Ndumu. The mosquitoes had been scrutinized as never before (though my eyes had failed to drop out as per Ken's orders). Besides the new species I had discovered, the catchers had brought in four kinds that were hitherto unknown from Natal province and four others new to the entire country. David Davis' mammal surveys and Bruce McIntosh's bird-netting program had thoroughly canvassed those two sections of the fauna. And from the original six virus isolates at Simbu Pan, the lab had brought the Tongaland list to thirteen. I must bring you up to date on the last four.

It was sometimes discouraging to suspect that one was handling virus-infected mosquitoes but unable to differentiate them from uninfected ones. Since I personally identified all the mosquitoes from Ndumu that the lab processed in 1959, I *know* in retrospect that I gazed upon at least three specimens that contained viruses new to the scientific medical world. It would have been ever so much more thrilling to have been able to spot them in the glass tube handed me by Dom-Dom or Magwalo. Even so, I was able to enjoy a vicarious feeling of discovery when the lab got around to announcing its triumphs.

Two of the viruses were very appropriately named: one, Usutu, for the local river, and the other Ndumu. Usutu came first, being derived from a batch of *Culex univittatus* collected on January 30, 1959. Its laboratory reactions showed it to be a Group B virus, such as we had sought but failed to find at Lumbo. It was apparently uncommon. Only one blood specimen out of 262 collected from cattle and sheep in South Africa reacted positively to the virus, while 198 specimens from human beings residing in South Africa and Moçambique were uniformly negative. However, the virus has since been recovered from *Taeniorhynchus* mosquitoes in Uganda, so it is not a unique Tongaland product.

Ndumu virus appeared in a sudden burst, on May 5, 1959, in a batch of *Aedes circumluteolus,* and two days later in a collection of *Taeniorhynchus uniformis.* Laboratory studies proved its relationship to an arbovirus family known as Group A, embracing viruses such as Chikungunya, which can cause severe disease in man. As yet it can be said only that Ndumu will infect human beings, but disease has not been reported. Immunity rates in various collections of blood specimens include five positive out of 66 tested from Moçambique, 12 out of 86 from Natal (in which Ndumu is situated) and five out of 89 from Bechuanaland. These could have been subclinical or asymptomatic infections. On the other hand, the individuals, who were chiefly natives living in the bush, may have been noticeably ill at the time of infection though fully recovered when their blood specimens were procured. Obviously Ndumu virus could cause fatal disease in the hinterlands without our knowledge.

One of the original Tongaland viruses—Simbu itself, after the name of the Simbu Pan expedition—had stood apart in the laboratory without evident relationship to other arboviruses. It might be said to have fallen into a group of its own, except for the difficulty of envisioning a group of one. However, its loneliness ended when Bruce isolated a second member of the Simbu family from a Spectacled Weaverbird that was caught in a net at Ndumu on May 15, 1959. He named it Ingwavuma virus after a small local tributary of the Usutu. Strange things were surely going on in the environment at that time, since the first mosquito infected with Ndumu virus had been collected only ten days previously. Of course we hadn't the faintest idea what sort of upheaval that may have been. Blood specimens from man, cattle and goats from the same region all reacted negatively with Ingwavuma virus. A couple of years after I left ABVRU, a second isolation was effected, this time from *Culex univittatus*. Since that mosquito feeds more readily on birds than some other kinds, infection of warm-blooded hosts by Ingwavuma virus may be confined to the avian world.

The thirteenth virus type from Tongaland was Sindbis, a Group A member already well studied in Egypt. Hence our data served merely to extend its known range. The lab secured three isolations —in January, May and October, 1959—two from *Culex univittatus* and one from *Taeniorhynchus africanus*, all collected at Ndumu and passing under my dazzled eye.

Not to belittle our work at ABVRU, it must be confessed never-theless that none of the newly discovered arboviruses was known to be an important cause of illness in human beings. The best—or worst—that could be said for most of them was that they were able to infect people, while a few such as Ingwavuma virus ap-parently couldn't even do that but had to pick on defenseless weaverbirds. Some critics were inclined to ask, "So what?"

In rising to defend arbovirus research, I found myself answer-ing some of my own self-criticism. All this work had been pioneer-ing, and we could not yet be ready to offer interpretations of its meanings or values. It is hard enough to direct your voyage of discovery without trying to write retrospective memoirs at the same time. We could ignore disparaging critics (though not en-lighten them). But someone more tolerant and of different scien-tific temperament might now take our novelties and elaborate them in realms that were foreign to our own talents. Or perhaps our work would stand as a curiosity for a long time without ap-parent usefulness. Yet there it was, soon to be published in full in medical and technical journals, freely available to those who might need it.

I wavered for quite a while between staying in Africa and going to Trinidad. As I have already said, the blight of apartheid even-tually forced my decision. There are times when it is not shameful to run away from distasteful things, one of those occasions being when you are a foreigner and consequently have no voice in the laws of a country. I must now tell of the last time I saw Jack at Ndumu, and perhaps you will understand me. That was the day the Emerald-spotted Wood Doves called their loudest and most urgently.

The head ranger did not stop by very often at our field head-quarters in the NRC camp, but when he did it was usually mid-morning and I would invariably invite him in for a cup of tea. Then we talked about fires, about the insect collection on the ceil-ing, about game and about "natives." The insects, by the way, I finally gave him for display at his field office where visitors checked in to see our hippos. Pottie told me in disgust one day that he happened to be on hand when a visitor complimented the ranger on his patience and industry in assembling such a collection. The ranger did not say in so many words that he had made the collection himself, but he did not bother to deny the visitor's implication that he had. Well, I didn't mind that—it seemed to bother Pottie more than me. Let the ranger take the credit if it would do him any good.

When Jack brought in the tea, the ranger addressed him sometimes in Zulu and at other times in English. When it was the latter language I would hear things such as, "Jack! You're as black as coal and you stink. Why don't you take a bath?" And much more. Pottie told me that the things said in Zulu were even worse.

The ranger thought this was very funny. He did it somewhat for my benefit, I'm sure, from the way he would smirk at me, but it was also part of his general form of treatment of all natives to "keep them in their places." Jack had to take it with a properly appreciative grin—that was part of the "treatment," too. (Though it has nothing to do with the case—apparently—I can state that Jack often bathed, and I'm sure he never stank as badly as I did when I came in from a sweaty day in the field.)

I used to talk to Pottie about it, and also lie awake thinking about it. The simplest thing would be not to ask the ranger for tea again, but the custom had now been established. In that case perhaps I should have asked him to stop speaking that way. However, I knew the ranger's reputation for being tough with the natives, and I could sense that I would offend him by saying anything. Since we were in the Ndumu Game Reserve as "guests" of the game department, or by their courtesy, as it were, I could not afford to antagonize the ranger and jeopardize ABVRU's entire program, for it would not take more than the ranger's report that we were conducting ourselves improperly to gain us an invitation to leave.

Yet every time such an episode was repeated, I felt myself further degraded in my own eyes, and I could not imagine what Jack must have thought of me for endlessly appearing to be party to the stale but hideously painful joke. So one day, after Jack had received the customary insults and then submissively withdrawn, I came out with it. The ranger was welcome to have tea under "my" roof, but while there he must observe the rules of my hospitality: either speak civilly to Jack or do not speak to him at all.

The ranger stayed quite a while that morning, talking about natives exclusively. The gist of his repetitive lesson was illustrated when my boys came in, their boxes filled with mosquito-containing tubes, and began to play soccer in the compound with a cheap rubber ball I had given them.

"Look at them," said the ranger. "Dirty, every one of them. And they'll all turn out badly, just as bad as Jack and the rest."

I couldn't see anything wrong or dirty about the boys. I called them over and recited my one Zulu sentence (laboriously learned from Pottie) to them. *"Ngi afuna izimiani emnengi"* (meaning, word for word, "I want little-flies many"). They doubled with laughter at me or at my accent or both, but with uproarious, enthusiastic, friendly gaiety.

"You see how nice they are," I said. "And just listen to their names! There's Dom-Dom, and that's Magwalo. Those are Msizwane, Ben, Sidoti, Hlweme, Siquza, Maguju, Majudi, Gebeza, Vimbi Izwe, Sdumu, Gadovu, Sele, Sdonolo, Mkumbuza and Magansela. One of those names—I forget which—means 'Strong man—he owns the world.' What's wrong with these boys? How are they any different from our own children?"

Apparently I really spoiled the idyll that morning. Pottie returned from Dupe's store a few days later with information that the ranger was spreading rumors about me. "He says you're some kind of Communist or something because you're sticking up for the natives."

He then trumped up a trespassing charge against Jack, who had been tolerated under his nose as a nontrespasser these several past years. On that pretext he arrived, on the day of the doves, with a strapping young assistant. They loaded Jack into their Land Rover and drove away.

I went to see him at police headquarters. His clothing was soiled and torn and one eye was badly damaged. They said Jack had tried to escape and that he had been injured while they re-strained him. I heard later that this was untrue. Just out of sight of the NRC compound they had stopped, hoping that he would try to get away, but when he remained docile they dragged him out and gave him a brutal beating on general principles.

He looked at me in a dazed manner but said nothing. For all I knew, he may have been thinking that it was all my fault, which indeed it was, for having spoken up for him in the first place. But of course there was nothing I could do or say now. I could not appeal to higher authority, for they were all against Jack, right up to the Prime Minister.

Jack was banished from Ndumu forever. He is said to have moved his family to the Simbu Pan region where he had relatives. The soil is not as rich as at Ndumu in that area, and consequently no one was likely to bother him there.

> *Ndon, ndon, ndon.*
> *Ngange né ngange zafa;*
> *Ngange né ngange bazeta;*
> *Man je inholizian yame*
> *ilokee yahlala iti*
> *Ndon, ndon, ndon, do, do*
> *Ndon, ndon.*

INDEX